Gender, Technology and Violence

Technological developments move at lightning pace and can bring with them new possibilities for social harm. This book brings together original empirical and theoretical work examining how digital technologies both create and sustain various forms of gendered violence and provide platforms for resistance and criminal justice intervention.

This edited collection is organised around two key themes of facilitation and resistance, with an emphasis through the whole collection on the development of a gendered interrogation of contemporary practices of technologically enabled or enhanced practices of violence.

Addressing a broad range of criminological issues such as intimate partner violence, rape and sexual assault, online sexual harassment, gendered political violence, online culture, cyberbullying and human trafficking and including a critical examination of the broader issue of feminist 'digilantism' and resistance to online sexual harassment, this book examines the ways in which new and emerging technologies facilitate new platforms for gendered violence as well as offering both formal and informal opportunities to prevent and/or respond to gendered violence.

Marie Segrave is an ARC DECRA Fellow researching unlawful migrant labour, exploitation and regulation. She is an Associate Professor in Criminology at Monash University and leads the Trafficking and Labour Exploitation research agenda of the Border Crossing Observatory (http://artsonline.monash.edu.au/thebordercrossingobservatory/) and is a researcher with the Monash Gender and Family Violence Focus Program. She researches migration, gender, violence and regulation in many forms.

Laura Vitis is a Lecturer in Criminology at the University of Liverpool in Singapore. Her research focuses on the regulation of and resistance to gendered violence facilitated by technology, youth sexting and the role of risk in the Sex Offender Register.

Routledge Studies in Crime and Society

For a full list of titles in this series, please visit www.routledge.com

23 Regulation and Social Control of Incivilities
Edited by Nina Peršak

24 Skinhead History, Identity, and Culture
Kevin Borgeson and Robin Valeri

25 Homicide, Gender and Responsibility
Edited by Sandra Walklate and Kate Fitz-Gibbon

26 Cybercrime Through an Interdisciplinary Lens
Edited by Thomas J. Holt

27 Domestic Violence in International Context
Edited by Diana Scharff Peterson and Julie A. Schroeder

28 Child Trafficking in the EU
Policing and Protecting Europe's Most Vulnerable
Pete Fussey and Paddy Rawlinson

29 Resettlement of Sex Offenders after Custody
Circles of Support and Accountability
David Thompson and Terry Thomas with Susanne Karstedt

30 Cybercrime and its Victims
Edited by Elena Martellozzo and Emma A. Jane

31 Money and the Governance of Punishment
A Genealogy of the Penal Fine
Patricia Faraldo Cabana

32 Gender, Technology and Violence
Edited by Marie Segrave and Laura Vitis

Gender, Technology and Violence

Edited by Marie Segrave
and Laura Vitis

Routledge
Taylor & Francis Group
LONDON AND NEW YORK

First published 2017 by Routledge

2 Park Square, Milton Park, Abingdon, Oxfordshire OX14 4RN
52 Vanderbilt Avenue, New York, NY 10017

Routledge is an imprint of the Taylor & Francis Group, an informa business

First issued in paperback 2019

British Library Cataloguing-in-Publication Data
A catalogue record for this book is available from the British Library

Library of Congress Cataloging-in-Publication Data
Names: Segrave, Marie, 1979– editor. | Vitis, Laura, editor.
Title: Gender, technology and violence / edited by Marie Segrave and Laura Vitis.
Description: 1 Edition. | New York : Routledge, 2017. | Series: Routledge studies in crime and society ; 31 | Includes bibliographical references and index.
Identifiers: LCCN 2016057649 | ISBN 9781138217232 (hardback) | ISBN 9781315441160 (ebook)
Subjects: LCSH: Women—Effect of technological innovations on. | Technology—Social aspects. | Technological innovations—Social aspects. | Women—Crimes against. | Sex role.
Classification: LCC HQ1233 .G464 2017 | DDC 305.4—dc23
LC record available at https://lccn.loc.gov/2016057649

ISBN: 978-1-138-21723-2 (hbk)
ISBN: 978-0-367-22709-8 (pbk)

Typeset in Times New Roman
by Apex CoVantage, LLC

Contents

Contributors

Gabrielle Bardall is an academic and an electoral assistance expert with a decade of experience supporting electoral processes in transitional states. She has worked in more than 25 countries with international organizations including UNDP, UN Women, DPKO, IFES, DRI, Carter Center and others. As a gender specialist with UNDP and UN Women in their New York headquarters and in country offices, she has contributed to several cutting-edge publications on enhancing women's political participation and has provided training on gender and elections to electoral administrators from more than 50 countries. Her research on gender and electoral violence pioneered the study of the field.

Kate Fitz-Gibbon is a Senior Lecturer in Criminology and an Honorary Research Fellow in the School of Law and Social Justice at University of Liverpool. Kate researches in the area of family violence, legal responses to lethal violence and the effects of homicide law and sentencing reform in Australian and international jurisdictions. Kate has published widely in the area, and her scholarship has received many accolades. One recent publication is (with Sandra Walklate, eds) (2016) *Homicide, Gender and Responsibility: An International Perspective*. Routledge: United Kingdom.

Fairleigh Gilmour is a lecturer at the University of Otago. She specialises in both criminology and gender studies. Her research interests include sex work governance; race, gender and class in media representations of crime; and the intersections of crime, addiction and mental health.

Emma A. Jane is an ARC DECRA Fellow and Senior Lecturer at the University of New South Wales. She investigates the ways in which the internet's rapid growth has delivered countless social benefits whilst also offering people new opportunities to attack each other with unprecedented venom and impunity. Her research covers the breadth of online misogyny, cyberbullying and digital mobs, with a focus on the social and ethical implications of emerging technologies.

JaneMaree Maher is a Professor of the Centre for Women's Studies & Gender Research, Sociology. Her program of research is focused in two key areas of gendered social science: women's work and family and gendered violence. Most recently she has co-authored (with Mason, McCulloch, Wickes & Pickering) *Prejudice Motivated Crime* (2017, Routledge).

Jude McCulloch is a Professor of Criminology in the School of Social Sciences at Monash University with degrees in law, commerce and criminology. She has published widely on the militarization of police, counter-terrorism laws and policing, and the politics of security and law and order, pre-crime and pre-emption and gendered violence. Her most recent book, with Dean Wilson, is *Pre-crime: Pre-emption, Precaution and the Future* (2016, Sage).

Sanja Milivojevic is a Senior Lecturer in Criminology at LaTrobe University, Australia. She has researched and published widely in the areas of human trafficking, gendered violence and technology. A recent publication is the monography (with Lee, Crofts & McGovern, 2015) *Sexting and Young People*, Palgrave McMillan.

Elena Pavan is an Assistant Professor in the Institute of Humanities and Social Sicences at Scuola Normale Superiore, Florence, Italy. Her interests are in the study of multi-actor political processes in global media and communication, especially in the internet governance domain, and in research methods, including social and semantic network analysis and lexicon-content analysis in social and political fields and in the study of transnational organized crime. One of her publications includes *Frames and Connections in the Governance of Global Communications: A Network Study of the Internet Governance Forum* (2012, Lexington Books).

Marie Segrave is an ARC DECRA Fellow researching unlawful migrant labour, exploitation and regulation. She is an Associate Professor in Criminology at Monash University and leads the Trafficking and Labour Exploitation research agenda of the Border Crossing Observatory (http://artsonline.monash.edu.au/thebordercrossingobservatory/). She researches migration, gender, violence and regulation in many forms.

Laura Vitis is a lecturer in Sociology, Social Policy and Criminology at Liverpool University (Singapore). Her research focuses on the regulation of and resistance to gendered violence facilitated by technology, youth sexting and the role of risk in the Sex Offender Register. With Fairleigh Gilmour, she most recently published in the journal *Crime, Media, Culture*.

Preface

Since the publication of Frances Power Cobbe's essay on 'Wife Torture in England' in 1878, the question of gendered violence has never been far from feminist concerns. Moreover, it is beyond dispute that the second-wave feminism of the late 1960s and 1970s put violence against women squarely on political, policy and academic agendas. In the intervening years, commentators have spoken and written about a 'third wave feminism' and more recently have observed the presence of a 'fourth-wave feminism'. In respect of the latter, Munro (2013) suggests that the increasing presence of the internet and its capacity for imaginative use have provided women with the space for a 'call-out culture'. She states:

> This culture is indicative of the continuing influence of the third wave, with its focus on micropolitics and challenging sexism and misogyny insofar as they appear in everyday rhetoric, advertising, film, television and literature, the media, and so on.
>
> (Munro 2013, p. 23)

Indeed, the internet has certainly provided the space for women to connect globally in ways not previously imagined. Moreover, it certainly provides an opportunity for engaging differently in campaign and research work (in this latter respect, see Carrington and Larkin 2016). At the same time, the ever-present nature of modern technology, without doubt an almost taken-for-granted feature of contemporary everyday life, carries with it both positive and negative consequences for all its users. The rapidity with which various media and their associated images and messages can now traverse the globe offers a stark example of both kinds of consequence. Indeed, the speed with which communication about events in one part of the world can take its toll in another can facilitate transparency on the one hand and misunderstanding on the other. The consequences associated with the publication of a cartoon of Mohammad in Denmark in 2005, sparking civil disturbances in the Middle East, are a good case in point. On the other hand, the use of mobile phone data to track down the bombers at the Boston Marathon in 2013 is a different, though no less pertinent, illustration of the different uses of modern technology. Such examples notwithstanding, it is also the case that modern technology has infiltrated our collective everyday life to the extent that it is now very difficult to imagine what life was like before Wi-Fi, mobile phones, the internet, social media and so on. Yet despite the evident grasp that such technology now has on our lives and the claims made for the space it affords fourth-wave feminism, that same technology also lends itself to behaviours long a focus of concerns voiced by second-wave feminism: gendered violence(s).

This edited collection fills a much-needed gap in our understanding of gendered violence(s) by exploring how technology can be and is used both to challenge such violence(s) and facilitate its (their) perpetuation. What is remarkable about this collection is that, while covering new ground, it points to the ongoing symmetry between the violence(s) of the past and those of the present, which have so preoccupied the feminist movement. A number of recurring themes are identified in this collection that are central to understanding the potential impact of this technologically focused world on women's lives. These are the nature of women's resistance(s) to violence(s), the assignation of responsibility for such violence(s), the subtle and not so subtle surveillance practices associated with such violence(s), the presumption of who is and who is not vulnerable to such violence(s) and the ways in which women are silenced about such violence(s). The continuities in these themes between women's lives of the past and those of the present require no further explication.

Every chapter in this collection adds a nuanced understanding of how each of these themes still frame women's lives, highlighting the ways in which the conventional boundaries among the public, the private and the virtual are increasingly blurred. Indeed, the lives documented here are constituted by an ongoing fuzzy interaction between all of these worlds, thus rendering women's use of technology a potential act of resistance while simultaneously making women more vulnerable. The use of technology to engage in minute-by-minute surveillance of all our lives (through location apps on mobile phones, for example) becomes more than monitoring mobile phone usage in the case of theft when put in the hands of a controlling partner. Yet at the same time, all of us are responsible and responsibilised through such apps – in this example, to protect our phones from fraudulent use. Unfortunately, those same processes when applied to understandings of gendered violence serve both to make the woman responsible and silence her abuse all at the same time. For those engaged in seeking to understand violence against women, in particular, such strategies of control are all too familiar – and represent more than 'call-out culture'.

Nonetheless, there are also opportunities here to challenge gendered violence(s). This was well illustrated in 2016 by the Stanford rape case, widely reported in social and other media, with the victim impact statement submitted to the court going viral (Willmott 2016). That the victim chose to remain anonymous in allowing her statement to be released added considerable power to her claim to speak for any woman who had undergone the process she had. The fact that the statement went viral endorses this claim, offering a window of opportunity for change. The potential here to use social media for what Ristovska (2016) has called 'strategic witnessing' is evident.

This edited collection opens the door for both criminology and victimology to make sense of what is happening in this technologically focused era, and to whom and how, in the spaces among the public, private and virtual. The behaviours under scrutiny are all too often left unquestioned simply because they are in plain sight. The chapters in this volume offer a pathway to making sense of the contemporary world and its connections with the past. As areas of investigation, both criminology and victimology are implicated in the politics of determining what are considered to be worthy topics of concern. It would be foolish to ignore the agenda implied by the collection presented here.

Sandra Walklate

References

Carrington, K and Larkin, A (2016), 'The politics of doing imaginative criminological research', in M Hviid-Jacobsen and S Walklate (eds), *Liquid Criminology*, Routledge-Ashgate, London, pp. 188–205.

Cobbe, F (1878), 'Wife-torture in England', *The Contemporary Review*, no. 32 (April 1878), pp. 55–87.

Munro, E (2013), 'Feminism: a fourth wave?', *Political Insight*, 22–25 September.

Ristovska, S (2016), 'Strategic witnessing in an age of video activism', *Media, Culture and Society*, DOI: 10. 1177/0163443716635866.

Willmott, D (2016), 'Is jury bias preventing justice for rape victims?', *The Conversation*, 9 June 2016, available at https://theconversation.com/is-jury-bias-preventing-justice-for-rape-victims-60090 (accessed 29 November 2016).

Acknowledgements

Marie and Laura extend their gratitude to the contributors, who have each brought unique insights into the interrogation of gender, technology and violence. This work, of course, is greatly indebted to our feminist forebears across many disciplines, who worked tirelessly to examine gender and to illuminate and value women's lived experience. This work as activism and advocacy within communities internationally and as scholarship within the academy is not complete, but it has gained much ground in recent decades.

We extend our thanks to our colleagues at Monash University and Liverpool University who have supported the development of this collection and, more broadly, offered an opportunity to enable our concerns and interests to be explored via research and publication.

We are indebted to the many leading scholars in this area – collectively the work being produced is shining a light on practices that can only be challenged by bringing them to bear.

Marie extends her personal thanks to her family, Steve, Sam and Thomas, and her Mum, whose support is essential, always. She also extends great thanks to Laura for her collegiality and her commitment and enthusiasm to transforming this kernel of an idea into a fully formed entity. The collection marks a beginning point for an interrogation of gender, violence and technology that must continue.

Laura extends her thanks to her family her Mum, Jonathon and Adam, who, although far away, are a never-ending source of support and guidance. She also thanks Marie for her tireless work, patience and scholarly support, without which this collection would not have been realised.

Finally, we are grateful to Routledge, in particular Thomas Sutton and Hannah Catterall, for believing in this project and offering administrative guidance and faith in the process. Last, but by no means least, we extend our gratitude to our copy editor extraordinaire, Julia Farrell, who is simply brilliant.

Abbreviations

APC	Association for Progressive Communications
BCDD	Broadband Commission for Digital Development
CALD	Culturally and Linguistically Diverse
CEDAW	*Convention on the Elimination of all Forms of Discrimination Against Women*
CEO	Chief Executive Officer
COE	Council of Europe
CSR	Corporate Social Responsibility
CSs	Community Standards (Facebook)
CSW	Commission on the Status of Women
DIG	Domain-Specific Insight Graphs
EOM	Election Monitoring and Observation Mission
GBPV	Gender-Based Political Violence
GBV	Gender-Based Violence
GDPV	Gender-Differentiated Political Violence
GPS	Global Positioning System
ICT	Information and Communications Technology
IGF	Internet Governance Forum
IRL	In Real Life
NGO	Non-Government Organisation
OECD	Organisation for Economic Co-operation and Development
OSCE	Organization for Security and Co-operation in Europe
SVWBS	Sexual Violence Won't Be Silenced
TFGV	Technologically Facilitated Gendered Violence
TFVAW	Technologically Facilitated Violence Against Women
ToS	Terms of Service
UK	United Kingdom
UN	United Nations
UNGA	United Nations General Assembly
UN.GIFT	United Nations Global Initiative to Fight Human Trafficking
US	United States
VAW	Violence Against Women
VAWG	Violence Against Women and Girls
VAWIP	Violence Against Women in Politics
VAWP	Violence Against Women in Politics

Introduction

Laura Vitis and Marie Segrave

Gender, Technology and Violence brings together original empirical and theoretical work examining how new digital technologies at once create and sustain various forms of gendered violence and provide platforms for resistance and criminal justice intervention. The volume offers an account at this global moment of the intersections of dominant discourses and policies around technology and the links between new and emerging technology and gendered violence. Over the past decade, violence facilitated by technology has come to the fore as a key international concern for law and policy makers, advocates and communities. There has been a sustained focus on technologically facilitated violence with respect to status issues, such as cyberbullying and young people (Chadwick 2014; Cross, Li and Smith 2012), cyberstalking (Yar 2013) and online 'sexual predators' (Barnes 2006; Cassell and Cramer 2008; Jenkins and boyd 2006; Shade 2007; Yar 2013). Yet there remains limited research literature that examines the role of technology in relation to gendered violence (Powell 2015) or the specific ways in which technology is used to both facilitate and respond to gendered violence. This volume aims to offer a unique and innovative contribution to the field, drawing on original empirical and theoretical interrogations.

Drawing from the work of international researchers, the collection provides reflections on and insights into a broad range of domestic and international criminal justice issues, such as intimate partner and family violence, rape and sexual assault, online sexual harassment, political and electoral violence and human trafficking. In doing so, this collection offers an examination of the ways in which new and emerging technologies facilitate new platforms for gendered violence or harassment and offer both formal and informal opportunities to prevent and/or respond to such violence. Importantly, each contribution critiques the assumptions surrounding 'technological mediation' and challenges the notion that technology is inherently risky and therefore a space where state (or non-state) interventions (via surveillance or control mechanisms are inherently protective (cf. Musto and boyd 2014, p. 5). At the heart of this collection is a feminist criminological concern to generate a more nuanced account of how gendered violence intersects with technology, how technology facilitates violence and how it is used to resist violence. This collection also reflects on the ways in which both state and non-state actors are engaging in strategies to counter gendered violence and the implications of these strategies. We turn now to explore the key themes of the collection.

Technology, harm and violence

In the advent of the digital revolution and the democratisation of the internet, many social practices and spheres of communication are now mediated by technology (Lupton 2015). So much so that it has been argued that such technology is 'becoming a constitutive part of what makes us human' (Miller and Horst 2012, p. 4). There are many important questions posed by these shifts. For criminologists, in particular, these shifts heralded changes to the landscape of crime and criminal justice. Early research into technology and crime focused on the first iterations of the internet, Web 1.0 (technologies including websites, emails, multiplayer games and dial-up internet – cf. Lupton 2015) and examined unlawful activity, including practices that have long existed but are accelerated or amplified via technology, such as money laundering or organised crime (trading in goods, arms or drugs). This field has been dominated by questions about national security, cyber security, financial crime, fraud and terrorism (Awan and Blackmore 2012; Yar 2013), as well as child pornography (Wall 2007; Yar 2013). These were seen as crimes that spanned the global sphere and necessitated analysis of the machinations of these technologies in order to understand the new frontier of crime and to identify when, where and how criminal justice systems ought to respond. However, the development of Web 2.0 (Wi-Fi, broadband internet access, wireless computing and social networking sites such as Facebook, Twitter and Instagram – cf. Lupton 2015) and the diversification of social interactions that it has enabled has reinvigorated explorations of technologically facilitated crime. These new developments have brought to bear questions about the role of the 'social' in technology use and the role of technology in new forms of interpersonal violence.

Within this period – from the emergence of the internet to the developments of Web 2.0 and beyond – critical reflections on digital technologies have emerged which recognise the necessity of taking into account the dialectical relationship between social hierarchies, gender relations, power structures and media landscapes and usage (Fuchs 2014). Early criminological research into technology and violence began to explore this relationship (between power and media usage) with regards to child pornography, child grooming (Jewkes and Wykes 2016; Yar 2013) and cyberbullying. Scholars examined the exploitation of children's vulnerability and limited power within online environments, which were subsequently positioned as causal factors for grooming (Jewkes and Wykes 2016, p. 946) and facilitators of child abuse and bullying (Cross, Li and Smith 2012). However, research is now refocusing on harms that illustrate the dialectic between gender and the networked techno-social landscape of the internet. Early predictions about the potential social implications of the internet purported that anonymity and the absence of traditional identity cues online could diminish social prejudices and the restrictions imposed by social hierarchies such as gender (Danet 1998; Herring 2003). Despite these early aspirations the fraught relationship between gender, power and media landscapes has been revealed through examinations of online hate speech (Shepherd et al. 2015; Yar 2013) and, more specifically, online gendered violence, misogyny and sexism (Citron 2009, 2014; Shepherd et al. 2015, p. 3), which illustrate that women are targeted for harassment in ways that yield gendered harms (Henry and Powell 2015). As such, new questions have begun to trouble criminologists whose research interests coalesce around the themes of crime and technology and the necessity for intersectional and feminist analysis.

An important point to make at this juncture pertains to the definition of technology in this context. This collection is not merely examining technologies as tools of (often gendered) violence – a limited position that belies the multifarious points of intersection between technology and violence. Rather, it takes a wider position. Underneath the umbrella term 'technology', we consider technological tools used to facilitate, accelerate and intensify violence and harassment (Citron 2014) and technologically facilitated spaces where cultural attitudes towards violence against women and cultures of harassment play out (Jane 2012). We also consider technologies as commodified entities with corporate and political architectures and policy frameworks that shape their usage and the violence and harassment that is enacted through them. Additionally, we position technology usage as a highly politicised phenomenon prone to engendering moral panics (cf. Cassell and Cramer 2008; Jewkes and Wykes 2016) and, as such, part of a slew of signal crimes ripe for law-and-order politics.

Technology-facilitated gendered violence

While the internet was imagined as a utopia that was accessible irrespective of geography, class or race (a largely unachievable utopia, though internet access globally continues to grow apace, from 1 per cent in 1995 to 40 per cent of the world's population having access in 2015–16 – World Bank 2016), online interaction remains underpinned and shaped by the same social dynamics that are manifested offline in ways that challenge the offline/online dichotomy (Shapiro 2013). As such, despite early aspirations of gender neutrality online, we are observing a period in which women are being targeted for harassment and rendered at risk of technologically facilitated violence. It is women who are often seen as at risk or who experience high volumes of online harassment and are targeted via new avenues of harassment and violence such as revenge pornography, doxing and online sexual harassment (cf. Henry and Powell 2014). These phenomena reinvigorate questions about the parameters of violence against women, the need to incorporate technologically facilitated practices in the continuum of violence against women (Henry and Powell 2015; Jane 2017) and the limitations placed on women's participation within the public sphere (Filipovic 2007). The inclusion of gender is, therefore, both purposive and responsive, particularly to the integration of technology into pre-existing forms of criminal violence against and harassment of women, the use of technology to manifest new forms of violence and the gendered impact of these new expressions.

As will be unpacked in the following chapters, technology has been used to facilitate a multifarious range of harassing behaviours targeting women, including but not limited to 'sexually based harms' (Henry and Powell 2016). These harms include online sexual harassment (Barak 2005; Henry and Powell 2014; Lindsay and Krysik 2012; Pew Research Centre 2014), revenge pornography (Citron 2009; Henry and Powell 2014), unsolicited nude images, online hate speech, rape threats, gender-based hate speech and online sexual solicitation (Baumgartner, Valkenburg and Peter 2010) – behaviours that are especially prominent in the lives of young women and girls (Barak 2005; Powell and Henry 2015) and women who are visible within the public sphere, particularly those who are openly feminist (Filipovic 2007; Jane 2012; Lewis, Rowe and Wiper 2016; Megarry 2014). Technology appears to be both a facilitator and an accelerant. For example, online hate speech and harassment have been mediated through the networked affordances of social media, such as the 'self-publishing, participatory, and "share' cultures"' of digital domains which enable those engaging in hate speech and misogyny to garner followers (Jane 2017, p. 3) and engage in mob attacks (Citron 2014). In addition, harassment is facilitated through the anonymity afforded by online spaces (Citron 2014), the potential for deindividuation (Siegel et al. 1986), disinhibition (Barak 2005; Bargh, McKenna and Fitzsimons 2002; Suler 2004), the inability to view the impact of our comments and practices (Citron 2014) and the broader cultural mores of the internet which are rooted in libertarianism (Jordan 2001) and a 'masculinist ethos' (Shepherd et al. 2015, p. 8). Moreover, in relation to intimate partner violence, technologies have been woven into established practices of such violence, including utilising GPS tracking in stalking and using social media to harass, monitor and control within violent relationships (Southworth et al. 2007). However, we situate this work as focused on the interaction of gender, technology and violence, not focused solely on technologically facilitated gendered violence (TFGV) or technologically facilitated violence against women (TFVAW), as it is often termed. As such, our work overlaps with the important work being undertaken in the field of TFGV but also seeks to expand the consideration of the contemporary interactions between technology, violence and gender, in accordance with our understanding of what constitutes technology as defined here.

We are interested in this collection not just in gendered *crimes* but also gendered violence and the gendering of responses to this violence. There is a wide range of gendered violence that intersects with technology but sits outside the bounds of the criminal law. This interpretation of violence as not only criminalised action (or inaction) broadens the scope of issues analysed in this collection, which are not focused on one crime type but rather on a wide range of contexts in which violence intersects with technology and gender. Moreover, these changes dovetail with the critical criminological call to shift the focus beyond criminalised harms and onto harms that occur outside the criminal law

(Hillyard et al. 2004), in addition to critiques of the limited analysis not only of harm but also of the gendered nature of harms enacted through technology (Henry and Powell 2015, p. 764). We are particularly interested in whether and to what extent we are witnessing the development of new forms of gendered violence that are enabled or enacted via technology or whether it is more useful to recognise emerging gendered violence as an extension of the entrenched practices of gendered harm. The issues we canvass in the collection include political violence, intimate partner violence, human trafficking, rape culture and online sexual harassment.

Technology as facilitator

The chapters within this collection expand on this question, exploring how technology is altering, intensifying and facilitating gendered violence. In Chapter 1, Maher et al. consider an issue of growing importance, technologically facilitated intimate partner violence. These authors draw from empirical research into women's experience of intimate partner violence in Victoria, noting that the use of texting, social media and tracking apps has created new spaces where men can enact violence against their intimate partners. This chapter also explores how technology extends the reach of violence by intensifying the potential for surveillance and control within women's social and work lives. In Chapter 3, Jane expands on this focus by detailing the nature, prevalence and implications of women's experiences of online harassment and hate speech, which she refers to as gendered cyberhate. In this chapter, Jane maps the range of behaviours that constitute gendered cyberhate to which the women in her study have been subjected – including explicit judgement on women's appearance or sexual attractiveness; ad hominem invective; hyperbolic misogyny, homophobia and/or sexually graphic imagery and rape threats; doxing; revenge pornography; and sextortion. This chapter also explores the ways in which these behaviours are now becoming more common and severe, as they are being perpetuated through mob attacks and sustained and repeated attacks. The chapter also provides a timely exploration of the psychological impacts of these forms of harassment.

In Chapter 6, Bardall shifts the focus substantially to examine the relationship between technology and a type of violence which has yet to be considered extensively in relation to TFGV – political and electoral violence against women. This chapter explores how the use of social media to threaten, demean, sexualise and trivialise women in politics draws upon pre-existing tropes and disadvantages women face within the political sphere in ways that further entrench barriers to women's political participation. The chapter also positions the control of women's access to technology as a new mechanism through which to deny access to political knowledge and avenues for political and civic engagement. In many ways, this chapter speaks to the failed aspirations of the internet as a tool of political and identity liberation and examines how, instead, the same challenges women face within offline political spheres are reified and manifested online.

In Chapter 4, Pavan provides a timely analysis of the range of roles that private corporations play in facilitating, regulating and responding to TFGV and harassment. Pavan examines how internet service providers, web-hosting services, search engines and participative networking platforms (such as Facebook, Twitter and YouTube) constitute actors that 'actively influence' the perpetration of gendered harms. Adopting an organisational perspective, Pavan explores how intermediaries hold a liminal position as 'hidden influentials' that facilitate harassment and abuse through their Terms of Service which enshrine free speech. This chapter brings to bear a much-excluded discussion in relation to technology, violence and harassment around the silent or hidden apparatuses that sustain abuse with limited culpability and scrutiny. Building upon themes established in Chapter 4, in Chapter 2, Milivojevic and Segrave also interrogate the role of institutions (and moral crusaders) in the manifestation of TFGV. They consider how the rapidly expanding anti-trafficking movement has shifted its focus onto e-trafficking, specifically the use of online technologies to facilitate trafficking. These authors assert that the political and policy focus on ICT as tools that have radically transformed sex trafficking is based on limited empirical support, and they address this oversight. Drawing upon a range of international trafficking policies, Milivojevic and Segrave critique embedded assumptions that there exists a trafficking–technology nexus. Specifically, this chapter examines how the political harnessing of e-trafficking has provided a platform through which to reinvigorate problematic constructions of women as vulnerable victims of sex trafficking, which conflate women with child victims. This chapter not only questions the claims that technology is a facilitator of human trafficking but also critiques how technology is being harnessed as a tool in the suite of international counter-trafficking efforts.

Technology as a site or tool of resistance

The second key theme explored in this collection is the complex relationship between technology as a tool or space of both facilitation and resistance. Within new media spaces exists potentialities or 'structural affordances' (boyd 2011, p. 46) of social networks such as persistence, replicability, scalability and searchability which are used as techniques to sustain TFGV; but they are also harnessed in resistance and response to harassment and other forms of online violence. This issue has been recently entertained in relation to digilantism (digital vigilantism). Scholars have observed that women are increasingly utilising social media to formulate networked counter-publics (Powell 2015; Salter 2013) that address, discuss and expose sexual violence (Powell 2015; Salter 2013), street harassment (Fileborn 2014) and online sexual harassment (Jane 2012; Megarry 2014; Vitis and Gilmour 2016) and engender support from online communities, particularly communities of feminist peers (Powell 2015). Moreover, in focusing on resistance, we reiterate our aim to examine technology as part of the social world. In Web 2.0, technology provides the tools through which violence is enacted, but also new public spaces or spheres within which violence unfolds. As Fraser (1990, p. 67) argues, public spheres do not just yield domination but also engender counter-publics or 'parallel discursive arenas' where subordinated groups 'invent and circulate counter discourses' and 'formulate oppositional interpretations of their identities, interests and needs'. As such, it is necessary to examine resistance as well as to critically engage with the emerging logics of what constitutes 'resistance' – which at times entail prevention and at other times post-harm responses.

Within this collection, extra-legal forms of resistance are critically considered in detail, with authors examining both the diverse potential that technology offers to foster resistance and prevent violence and the problematic implications embedded within these approaches. In Chapter 3, Jane provides a detailed account of the range of individualised tactics women use to resist gendered cyberhate such as blocking objectionable content, engaging harassers personally and naming and shaming harassers on social media. Drawing from two case studies – of Australian journalist Tracy Spicer and academic Jenna Price – Jane maps the rationales for and manifestations of the different types of resistance these women deploy as targets of cyberhate. The author argues that the impetus for and celebration of such digilante responses illustrates a responsibilisation and reprivatisation of violence and that if 'women are finding a range of innovative ways to dodge, de-fang and/or declare guerrilla warfare against misogyny online', this 'should never be used, either explicitly or implicitly, to relieve regulators of their responsibilities with regard to cyber violence'.

Chapter 5 extends on the discussion laid out by Jane and brings to bear questions about the utilisation of social media to foster resistance to sexual violence and challenge dominant rape cultures. In this chapter, Gilmour and Vitis examine young men's utilisation of Twitter and YouTube to generate counter-publics that question the victim-blaming attitudes evident in response to the Steubenville rape case. The authors texturise the pre-existing literature on social media as a tool for resistance to rape narratives by exploring how such challenges are manifesting outside feminist circles and being generated by young men. This chapter considers the role of masculinity within these counter-publics and the harnessing of new media to facilitate these messages of resistance.

This collection is committed to the idea that there are opportunities to challenge gendered violence, directly and indirectly, via technology. However, we draw on a long feminist critical criminological history of raising questions about the extent to which well-meaning efforts to 'protect' translate into approaches that reify stereotypical constructions of women's position and dependency as victims (Bulmiller 2008). This particular reflection is brought to bear in Chapter 2, in which Milivojevic and Segrave argue that perceptions of technology – specifically, its perceived 'unregulated and deterritorialised nature' – have fostered problematic solutions to e-trafficking that are aligned with law-and-order frameworks. This chapter explores how, as part of efforts to combat e-trafficking, ICTs have been harnessed and deployed to increase surveillance of women's online activities via growing requests for data retention by governments, online censorship and surveillance of classified and adult-services websites. Milivojevic and Segrave consider these approaches as methods that extend the reification of women as solely victims of trafficking and construct barriers to women's mobility from the

Global South to the Global North. The problems associated with targeting and tackling gendered violence online are also considered in Chapter 4, where Pavan interrogates the limited ways in which internet intermediaries take responsibility for the harassment and abuse occurring on their platforms. This chapter considers how intermediaries function as 'free agents' without specific normative obligations to manage or measure their responsibilities to victims. Pavan argues that, absent such obligations, intermediaries hold a position of impunity with respect to responding to the rights violations of their users.

Responsibility, technology and violence

The final theme that runs through this collection is that of responsibility. In acknowledging that technology is used in practices of resistance and as a technique of responding to violence and harassment, broader questions are raised about where responsibility for TFGV lies or is being fostered. What is made evident within this collection is that, as technology melts into everyday life and communication (Lupton 2015), the range of parties that come to assume positions of responsibility for TFGV diversifies. Where offline violence focuses on the victim–offender–state triad, online violence brings individuals, markets, global communities, social activists and state institutions into the sphere of responsibility in new configurations that decentre this triad.

Individual responsibilisation

One of the key ways in which this question of responsibility is brought to the fore is in relation to individual responsibility. Critical reflections on the privatisation of responsibility for TFGV first emerged within analyses of state-sponsored responses to teen sexting (Albury and Crawford 2012). Sexting educational campaigns across the United Kingdom and Australia employed moralising narratives that blamed young victims of non-consensual sexting for taking images in the first place, encouraging them to take responsibility for their future victimisation by limiting their digital citizenship and avoiding nude image-taking and sharing (Albury and Crawford 2012; Karaian 2012; Ringrose and Renold 2012). Questions of responsibilisation in relation to TFGV have since transformed from these early panics of youth sexting and re-emerged to encompass the new complexities inherent to individual responses and resistance to TFGV. Recent international survey research on the nature, impact and responses to online gendered harassment indicates that women are utilising a range of individual responses to manage harassment, including confronting perpetrators, using security measures like blocking, using security settings to filter abuse and diminishing their online presence by changing usernames or desisting from social media use (Lewis, Rowe and Wiper 2016; Pew 2014; Powell and Henry 2015). Within this broader tapestry of individual responses exists the uptake of digilantism, which includes exposing and shaming harassment and harassers on social media. As Harris (2004) argues, post-feminist discourses (within which practices of digilantism are situated) align with neoliberal political rationalities and technologies that engender active prudential citizens. This point is considered within this collection. In Chapter 3, Jane raises questions as to whether such individual practices of digilantism should be celebrated as resistance and not seen as a manifestation of the problem. The author engages critically with the liberatory potential of individual and collective practices of digilantism and argues that such responses should be considered as responses to oppressive circumstances rather than solutions to the problems of gendered cyberhate. As such, Jane cautions that to focus on individual strategies and responsibilities deviates attention from the necessity for robust legislative support for victims of TFGV and harassment. A similar theme is addressed in Chapter 1. In their examination of the growth of apps and websites aimed at educating women on technology and intimate partner violence and providing women with the tools to prevent and manage violence within their relationships, Maher, McCulloch and Fitz-Gibbon question the latent function of these tools as a reprivatisation of violence which further responsibilises women to manage their own safety in the absence of state intervention or assistance. While these chapters ask important questions about the problems engendered by the responsibilisation of individual citizens, particularly female citizens, others turn their focus to the responsibilities of institutions, particularly internet intermediaries, and the limited responsibility they bear for managing, reporting and controlling those who use their services to perpetrate harassment.

Service providers

Service providers like Twitter, Facebook and Instagram (platforms where harassment proliferates) are currently being interpellated as key actors in the broader network of responsibilised parties for preventing TFGV. Website operators and their moderators are argued to be 'important sources of deterrence and remedy' by monitoring and removing harassing content, suspending user accounts (Citron 2014, p. 167) and setting clear community conduct and service standards (Powell and Henry 2015, p. 4). While such arguments position service providers as responsible for preventing harassment, media scholars have argued that the platform and algorithmic politics of social media sites (such as the anonymity of users) are in fact enabling harassment by protecting perpetrators and increasing the potential for harassment to be enacted (Bucher 2012; Gillespie 2015; Massanari 2015). Despite this, the position adopted by some site operators reflects libertarian and utilitarian principles that place sole responsibility for abuse on the user rather than the service provider (Citron 2014, p. 168). This attitude was evident in the statement of former chief executive officer (CEO) of Reddit, Yishan Wong, justifying Reddit's stance on hosting non-consensually obtained nude images of American actress Jennifer Lawrence, in which he claimed that it is the 'user's responsibility to choose between good and evil' (Isaac 2014, para 9). Moreover, questions have been raised as to whether social media providers in fact benefit from harassment occurring on their sites. For example, Massanari (2015, p. 8) argues that Reddit's initial reluctance to ban Lawrence's nude images 'may have been monetary: in 6 days, subscribers purchased enough Reddit gold (a kind of currency that defrays Reddit's server costs) to run the entire site for a month' – thus illustrating that abuse occurs within a 'recursive relationship between technological design', social patterns and the social geographies of its users (Salter 2017, p. 698). Despite such observations, poor moderation and limited attempts to enforce anti-harassment policies remain (Massanari 2015; Salter 2017). However, within the public sphere, service providers have begun to address these issues. For example, Twitter CEO Jack Dorsey recently admitted that the platform had failed to protect users from abuse, and after this admission, Twitter's director of product management and user safety indicated that the platform was improving their reporting, blocking and enforcement procedures for abusive content and accounts (Hern 2015). Similarly, Reddit has employed an anti-harassment policy (Salter 2017), and Instagram has deployed comment filters (allowing the user to list the words that they would like to be blocked from comments) and personalised block lists (Tsukayama 2016). The responsibility of intermediaries to prevent abuse and harassment is explored in detail in Chapter 4. In this chapter, Pavan argues that internet intermediaries have remained in the background of discussions about strategies to prevent TFGV, largely abdicating social responsibility for preventing harassment. However, Pavan examines how this absence of responsibility is also being challenged by civil society initiatives which are exposing and highlighting their integral roles in both facilitating and preventing abuse. This unsettles the construction of service providers as apolitical communication 'hosts' and repositions corporate entities as responsible parties.

Community responsibility

This collection also raises questions about the shifting role of community responsibility. As Powell and Henry (2015, p. 5) argue, 'ultimately, we are all members of digital society and we all have a role to play in ensuring that our social experiences are equal and inclusive – whether occurring online or offline'. Digital formulations of communal responsibility are evident in practices of resistance that utilise social media to expose sexual violence, online sexual harassment and rape-supportive cultures through social networks, particularly those that encourage other social media users to bear witness to harassment. In bearing witness, victims of violence place responsibility on the community to acknowledge and engage with the realities of violence against women. As Tait (2011, p. 1220) argues, 'bearing witness extends beyond seeing through practices of enacting responsibility'. The role of community responsibility is examined in Chapters 5 and 1. Gilmour and Vitis explore the role of networked communities in configuring new global communities of practice that position community members as responsible for intervening in sexual violence and contesting rape narratives. Moreover, in Chapter 1, Maher et al. draw from Judith Butler's concepts of relationality and collective responsibility to argue that it is necessary to deviate from further privatising intimate partner violence through technology and to advocate for social and political responsibility.

Conclusion

As is clear from this overview, this collection draws on a range of interdisciplinary approaches to the interrogation of the violence–technology–gender nexus. The chapters in this collection do not offer an exhaustive account of these intersections, but they do offer a varied and considered conversation about these issues. We believe that each of the chapters offers an account of technology, gender and violence that challenges some of the entrenched views of what gendered violence 'looks like' and what is required to curtail it. In some ways, the collection is disheartening, as it reveals the persistence of gendered violence and, often, accounts of violence that lead to simplistic and problematic narratives of responsibility. However, the chapters also provide critiques that serve as a counterpoint to dominant narratives and, in various ways, offer a critical feminist account of how we can move forward.

References

Albury, K and Crawford, K (2012), 'Sexting, consent and young people's ethics: beyond Megan's story', *Continuum*, vol. 26, no. 3, pp. 463–73.

Awan, I and Blackmore, B (2012), *Policing Cyberhate, Cyber Threats and Cyber Terrorism*, Ashgate, Surrey.

Barak, A (2005), 'Sexual harassment on the internet', *Social Science Computer Review*, vol. 23, no. 1, pp. 77–92.

Bargh, JA, McKenna, KYA and Fitzsimons, GM (2002), 'Can you see the real me? Activation and expression of the "true self" on the internet', *Journal of Social Issues*, vol. 58, no. 1, pp. 33–48.

Barnes, SB (2006), 'A privacy paradox: social networking in the United States', *First Monday*, vol. 11, no. 9, available at www.firstmonday.org/ISSUES/issue11_9/barnes/ (accessed 28 October 2012).

Baumgartner, S, Valkenburg, P and Peter, J (2010), 'Unwanted online sexual solicitation and risky sexual online behaviour across the lifespan', *Journal of Applied Developmental Psychology*, vol. 31, no. 6, pp. 439–47.

boyd, d (2011), 'Social network sites as networked publics: affordance, dynamics and implications', in Z Papacharissi (ed.), *Networked Self: Identity, Community, and Culture on Social Network Sites*, Taylor & Francis, New York, pp. 39–58.

Bucher, T (2012), 'Want to be on the top? Algorithmic power and the threat of invisibility on Facebook', *New Media & Society*, vol. 14, pp. 1164–80.

Cassell, J and Cramer, M (2008), 'High tech or high risk: moral panics about girls online', in T McPherson (ed.), *Digital Youth, Innovation, and the Unexpected*, MIT Press, Cambridge, pp. 53–75.

Chadwick, S (2014), *Impacts of Cyberbullying: Building Social and Emotional Resilience in Schools*, Springer, Cham.

Citron, D (2009), 'Law's expressive value in combating cyber gender harassment', *Michigan Law Review*, vol. 108, pp. 373–416.

Citron, D (2014), *Hate Crimes in Cyberspace*, Harvard University Press, Cambridge.

Cross, D, Li, Q and Smith, P (2012), *Cyberbullying in the Global Playground: Research From International Perspectives*, Wiley-Blackwell, Chichester, Malden, MA.

Danet, B (1998), 'Text as mask: gender and identity on the Internet', in S Jones (ed.), *Cybersociety* 2.0, Sage, Thousand Oaks, CA, pp. 129–58.

Fileborn, B (2014), 'Online activism and street harassment: digital justice or shouting into the ether', *Griffith Journal of Law & Human Dignity*, vol. 2, no. 1, pp. 32–51.

Filipovic, J (2007), 'Blogging while female: how internet misogyny parallels "real-world" harassment', *Yale Journal of Law and Feminism*, vol. 19, no. 1, pp. 295–303.

Fraser, N (1990), 'Rethinking the public sphere: a contribution to the critique of existing democracy', *Social Text*, vol. 25/26, pp. 56–80.

Fuchs, C (2014), *Social Media: A Critical Introduction*, Sage, London and Thousand Oaks, CA.

Gillespie, A (2015), *Cybercrime: Key Issues and Debates*, Taylor & Francis, London.

Harris, A (2004), *Future Girl: Young Women in the Twenty-First Century*, Routledge, New York.

Henry, N and Powell, A (2014), *Approaches to Overcoming a Rape Culture*, Palgrave Macmillan, London.

Henry, N and Powell, A (2015), 'Embodied harms: gender, shame, and technology-facilitated sexual violence', *Violence Against Women*, vol. 21, no. 6, pp. 758–79.

Henry, N and Powell, A (2016), 'Technology-facilitated sexual violence: a literature review of empirical research', *Trauma, Violence, & Abuse*, Online First, pp. 1–14.

Hern, A (2015), 'Twitter CEO: we suck at dealing with trolls and abuse', *The Guardian*, 5 February, available at www.theguardian.com/technology/2015/feb/05/twitter-ceo-we-suck-dealing-with-trolls-abuse (accessed 10 May 2016).

Herring, SC (2003), 'Gender and power in online communication', in J Holmes and M Meyerhoff (eds), *The Handbook of Language and Gender*, Blackwell Publishers, Oxford, pp. 202–28.

Hillyard, P, Pantazis, C, Thombs, S and Gordon, D (eds) (2004), *Beyond Criminology, Taking Harm Seriously*, Pluto Press and Fernwood Publishing, London.

Isaac, M (2014), 'Reddit and 4chan begin to button up', *The New York Times blog*, available at http://bits.blogs.nytimes.com/2014/09/08/reddit-and-4chan-begin-to-button-up/?_r=0 (accessed 20 January 2016).

Jane, E (2012), 'Your a ugly, whorish, slut', *Feminist Media Studies*, vol. 14, pp. 531–46.

Jane, E (2017), *Misogyny Online: A Short and Brutish History*, Sage, London.

Jenkins, H and boyd, d (2006), *Discussion: MySpace and Deleting Online Predators Act, Digital Divide Network*, available at www.webcitation.org/5JcDJkn4u (accessed 10 February 2014).

Jewkes, Y and Wykes, M (2016), 'Reconstructing the sexual abuse of children: "cyber-paeds", panic and power', *Sexualities*, vol. 15, no. 8, pp. 934–52.

Jordan, T (2001), 'Language and libertarianism: the politics of cyberculture and the culture of cyberpolitics', *The Sociological Review*, vol. 49, no. 1, pp. 1–17.

Karaian, L (2012), 'Lolita speaks: "sexting," teenage girls and the law', *Crime, Media, Culture*, vol. 8, no. 1, pp. 57–73.

Lewis, R, Rowe, M and Wiper, C (2016), 'Online abuse of feminists as an emerging form of violence against women and girls', *British Journal of Criminology*, Online First, pp. 1–20.

Lindsay, M and Krysik, J (2012), 'Online harassment among college students: a replication incorporating new internet trends', *Information, Communication & Society, Special Issue: A Decade in Internet Time – The Dynamics of the Internet and Society*, vol. 15, no. 5, pp. 703–19.

Lupton, D (2015), *Digital Sociology*, Routledge, Abingdon, Oxon.

Massanari, A (2015), '#Gamergate and the Fappening: how Reddit's algorithm, governance, and culture support toxic technocultures', *New Media & Society*, Online First, pp. 1–18.

Megarry, J (2014), 'Online incivility or sexual harassment? Conceptualising women's experiences in the digital age', *Women's Studies International Forum*, vol. 47, pp. 46–55.

Miller, D and Horst, H (2012), 'The digital and the human: a prospectus for digital anthropology', in H Horst and D Miller (eds), *Digital Anthropology*, Berg, London, pp. 3–35.

Pew Research Centre (2014), *Online Harassment*, available at www.pewInternet.org/files/2014/10/ PI_OnlineHarassment_72815.pdf (accessed 2 January 2015).

Powell, A (2010), *Sex, Power and Consent: Youth Culture and the Unwritten Rules*, Cambridge University Press, Port Melbourne.

Powell, A (2015), 'Seeking informal justice online: vigilantism, activism and resisting a rape culture in cyberspace' in Anastasia Powell, Nicola Henry, Asher Flynn (eds), *Rape Justice: Beyond the Criminal Law*, Palgrave Macmillan, Hampshire, United Kingdom, pp. 218–237.

Powell, A and Henry, N (2015), *Digital Harassment and Abuse of Adult Australians: Digital Harassment and Abuse*, RMIT University, Melbourne.

Ringrose, J and Renold, E (2012), 'Slut-shaming, girl power and "sexualisation": thinking through the politics of the international SlutWalks with teen girls', *Gender and Education*, vol. 24, no. 3, pp. 333–43.

Salter, M (2013), 'Justice and revenge in online counter-publics: emerging responses to sexual violence in the age of social media', *Crime, Media, Culture*, vol. 9, no. 3, pp. 225–42.

Salter, M (2017), *Crime, Justice and Social Media*, Routledge, London.

Shade LR (2007), 'Contested spaces: protecting or inhibiting girls online?', in S Weber and S Dixon (eds), *Growing Up Online: Young People and Digital Technologies*, Palgrave Macmillan, New York, pp. 227–44.

Shapiro, E (2013), *Gender Circuits: Bodies and Identities in a Technological Age* (2nd edn), Routledge, London.

Shepherd, T, Harvey, A, Jordan, T, Srauy, S and Miltner, K (2015), 'Histories of hating', *Social Media + Society*, vol. 1, no. 2, pp. 1–10.

Siegel, J, Dubrovsky, V, Kiesler, S and McGuire, T (1986), 'Group processes in computer-mediated communication', *Organizational Behaviour and Human Decision Processes*, vol. 37, no. 2, pp. 157–87.

Southworth, C, Finn, J, Dawson, S, Fraser, C and Tucker, S (2007), 'Intimate partner violence, technology, and stalking', *Violence Against Women*, vol. 13, no. 8, pp. 842–56.

Suler, J (2004), 'The online disinhibition effect', *CyberPsychology & Behavior*, vol. 7, no. 3, pp. 321–6.

Tait, S (2011), 'Bearing witness, journalism and moral responsibility', *Media, Culture & Society*, vol. 33, no. 8, pp. 1220–35.

Tsukayama, H (2016), 'Instagram rolls out its comment filter to everyone', *The Washington Post*, 12 September, available at www.washingtonpost.com/news/the-switch/wp/2016/09/12/instagram-rolls-out-its-comment-filter-to-everyone/ (accessed 30 September 2016).

Vitis, L and Gilmour, F (2016), 'Dick pics on blast: a woman's resistance to online sexual harassment using humour, art and Instagram', *Crime, Media, Culture*, Online First, pp. 1–21.

Wall, D (2007), *Cybercrime: The Transformation of Crime in the Information Age*, Polity Press, London.

World Bank (2016), *Internet Users*, available at http://data.worldbank.org/indicator/IT.NET.USER.P2?page=6&cid=GPD_44 (accessed 17 November 2016).

Yar, M (2013), *Cybercrime and Society: Crime and Punishment in the Information Age* (2nd edn), Sage, London.

1 New forms of gendered surveillance? Intersections of technology and family violence

JaneMaree Maher, Jude McCulloch and Kate Fitz-Gibbon[1]

The late twentieth and early twenty-first centuries saw considerable shifts in discourses addressing and relating to family violence. Feminist challenges to the previous invisibility, acceptability and privatisation of such violence have driven social change: responsibility to respond to, and prevent, family or domestic violence, most particularly violence against women and children, is increasingly accepted across developed and developing nations as both a national and an international obligation. The desired outcomes of these obligations are as yet unrealised; the scope and scale of family violence, of which the most prevalent form is intimate (ex)partner violence by men against women, are still emerging (Garcia-Moreno et al. 2006; Hoeffler and Fearon 2015). The work of preventing such violence and enhancing security for women and children has begun but is far from complete. In this period of increased recognition and policy reform, the emergence of new digital technologies has fundamentally changed social interactions, presenting new forms of visibility, communication and connection. Access to knowledge is made easier, even as forms and repositories of knowledge proliferate and create new types of risk and new opportunities for crime (Grabosky 2016). In this chapter, we examine the meaning and impacts of these new technologies in the terrain of family violence. Our interest here is in how such technologies interact with changing social and political responses to family violence. What new forms of responsibility are engendered by technologies that both create and respond to risk? How might enduring myths about family violence and responses to family violence that create stigma and risk for women be reinvigorated by digital technologies?

While recognising the opportunities for enhanced security from family violence and the positive social change that are enabled by new technologies, we examine the simultaneous potential for harms from these same technologies. Using Judith Butler's account of the profound interconnections of vulnerability and violence in *Precarious Life: The Powers of Mourning and Violence* (2004), we argue that these technologies may reinvigorate critical normative aspects of family violence discourses that have proved difficult to shift. Men's capacity to perpetrate violence against intimate partners is facilitated by new digital technologies that significantly expand opportunities for surveillance and harassment. Discourses that continue to assume that women are responsible for securing their own safety from family violence are extended; the use of digital communication is at once seen as an obligation in escaping violence while also creating significant new risks. Rethinking notions of privacy and security has been central to challenging norms around the meaning and acceptability of family violence, and privacy and security are also central to social responses to the rise of new digital technologies. Yet the meanings and implications of such terms in the context of family violence are complex. 'Privacy is complex and does not have a universal definition. . . . Privacy is negotiated in response to circumstances, rather than an enforcement of rules' (Dimond, Fiesler and Bruckman 2011, p. 415). How we understand and respond to the differing and shifting implications of privacy in technology-facilitated family violence abuse, as well as how we understand and assign responsibility in relation to this type of violence, will have significant impacts on the safety of women and children. We argue that these technologies may work to create new forms of surveillance of women's activities and responsibilities in terms of family violence, such that women are once again held responsible for securing their own safety and encouraged to stay silent (this time online) as part of this responsibility.

In this chapter, we look first at how new digital technologies facilitate violence against women, with a particular focus on how technological surveillance creates new possibilities for the commission of family violence. We then review the opportunities for enhanced security and safety that are created by these same new technologies and consider how discourses about women's responsibility to secure their own safety operate in this context. We argue that these technologies have the potential to reprivatise family violence and reinvigorate notions of women's, rather than social or community, responsibility to respond to and prevent such violence.

This chapter draws from our recent research examining risk assessment and management practices in Victoria, Australia (see McCulloch et al. 2016). Here we draw specifically on the victim-survivor interview and focus group data collected as part of that research, as it relates to technology-facilitated family violence, and how these technologies form part of the family violence some women experience. Interviews and focus groups were conducted with 24 victim-survivors. All participants were in contact with and were recruited through a Victorian family violence support service. The interviews focused on women's experiences of risk assessments and service responses across Victorian metropolitan and regional areas. Around half of the women participants had left the abusive relationship more than five years prior, while 12 of the women had left the abusive relationship within the 18 months prior to the interview/focus group (for further details, see McCulloch et al. 2016). To ensure participant confidentiality, all victim-survivors are referred to throughout this chapter by a pseudonym accompanied by a brief descriptor to assist in contextualising their experience.

Butler (2004) suggests that we are connected by shared vulnerabilities and shared complicity in violence. Such connections need to be carefully considered in light of the way in which digital technology presents both a promise and a threat in the realm of women's security and family violence. In particular, the opportunities offered by technologies cannot be assumed to make women more responsible for securing their own safety. We take up Butler's argument that concepts of autonomy are necessary to achieve 'legal protections and entitlements' (2004, p. 25) but insufficient to address our 'collective responsibility' (2004, p. 44) to each other. It is critical that digital technologies in the realm of family violence are considered with attention to shared social and political obligations to combat and prevent violence against women and children.

Chancy connections: intimacy, technologies and surveillance

> The technology is new to many advocates and victims, necessitating that advocates learn about and address these high-tech tactics, but always in the larger context of a victim's stalking experience. The rapid expansion and availability of new information technologies poses new threats to both victims and domestic violence service providers.
>
> (Southworth et al. 2007, p. 843)

As Levy (2014) has observed, the rise of communication technologies and social media platforms has profoundly influenced the life course of intimate relationships. Apps may inaugurate relationships, but even if they do not, social media presence now forms part of the relational information that people gather about each other. A wide range of communication technologies, most notably mobile phones, operate as key relationship tools (Burke et al. 2011). The amount of data available online about each person has expanded exponentially, as have points of connection to digital data:

> From women's bodies and cycles to their whereabouts, communications, and activities, services from Glow to Wife Spy to Girls Around Me expose women especially to data collection, invasive monitoring, and increased visibility. Intimate surveillance gives us a sense of control over a fundamentally uncontrollable dimension of personal life: we can only control that which we can track and measure.
>
> (Levy 2014, p. 688)

Levy (2014) argues that such changes have altered the contemporary boundaries and limits of 'intimate surveillance'; concepts of privacy and what are deemed appropriate forms of contact and visibility in relationships are developed in relation to these new expectations of visibility and access. Social media presence means that intimate relationships are developed through knowledge that is proffered by the person and simultaneously gleaned through online platforms. Henry and Powell (2015, p. 114) argue that this change has been significant in how practices of intimacy and relationships are understood: as a consequence of these technological developments, 'the boundaries of acceptable "romantic" behaviours and gender-based violence have also broadened and overlapped – particularly in relation to online harassment and stalking behaviours'. Burke et al. (2011) similarly suggest that the lines of 'appropriate and intrusive' communication and contact in relationships are now blurred and uncertain: multiple text messages at the beginning of a relationship may be romantic but threatening and improper as it ends.

In the last two decades, there has been increasing recognition of the ways in which digital technologies and the forms of social exchange they facilitate are being used in what Woodlock (2015) identifies as 'technology-facilitated violence'. Henry and Powell (2016) use the term 'technology-facilitated sexual violence' to capture many of the same practices of technological abuse carried out by partners and ex-partners (as well as the use of technology in a range of other forms of violence). Southworth et al. (2007) prefer to use the term 'stalking with technology' (2007, p. 844) to reinforce the point that new technologies extend the reach and nature of pre-existing forms of violent abuse against women.

Technology enables many practices that constitute forms of family violence. These practices include:

- threatening or abusive phone calls
- repetitive threatening or abusive text messages and emails
- checking or hacking email accounts
- monitoring internet use
- 'revenge porn', whereby a person distributes or posts false, humiliating, intimate or sexualised videos or photos without the other person's consent
- harassing or threatening the victim or the victim's friends and family on social networking sites such as Facebook and Twitter, or dating, chat and games sites
- tracking the victim's location through apps and 'find my phone' services
- smartphone spyware. (DVRCV 2015, p. 127)

There has been considerable discussion about whether these forms of abuse are best regarded as new forms of family violence or whether they should be understood as extensions of stalking (Mason and Magnet 2011; Southworth et al. 2007; Yar 2013, pp. 128–36). Stalking, a term that first came to prominence in the 1980s, describes 'persistent harassment in which one person repeatedly imposes on another unwanted communications and/or contacts' (Mullen, Pathé and Purcell 2001, p. 9). Mason and Magnet consider that 'stalking is not a new phenomenon. And yet, new technologies complicate how women experience violence as well as how they are able to protect themselves' (2011, p. 107). Whatever decision is made about the best framework to understand and interpret this form of family violence, it is clear that the reach and impacts of technology-facilitated family violence are profound.

Recent research in the United States, Australia and the United Kingdom has begun to reveal the scope and nature of 'technologically facilitated abuse', although Henry and Powell (2016) argue that there is still very limited research on this type of abuse when perpetrated against adult women rather than adolescents or young adults. While robust empirical data is still emerging about the extent and prevalence of this type of family violence, existing research indicates that such abuse is rapidly increasing (George and Harris 2014; Henry and Powell

2016; McKee 2006; Woodlock 2016). The studies and reports that have emerged indicate that the use of abusive text messaging is prevalent (Cavezza and McEwan 2014; DVRCV 2015; Royal Commission into Family Violence [RCFV] 2016; Woodlock 2016) – frequency, persistence and threatening content all operate to damage the recipient's sense of safety and security and to create a sense of constant surveillance (Woodlock 2015). As Woodlock (2016) further identifies, perpetrators may use coded messages or patterns of delivery that are targeted particularly at their victims but may not be readily understood by others as threatening or damaging. Women victims may be reluctant to reveal this type of abuse, fearing that such revelation may lead to escalation (Dimond, Fiesler and Bruckman 2011) or be seen as trivial. A police officer in our recent study described an incident in which a woman indicated to him that her ex-partner was breaching an intervention order by texting her 'a few times': when police obtained the phone records, there were more than 350 texts in a period of two months. Even when text messages are clearly threatening and in breach of intervention orders, research suggests that it may be difficult to attract any police attention or response (Cavezza and McEwan 2014; Dimond, Fiesler and Bruckman 2011). As one victim explained:

> I have got like 50 messages from him that are breaches, saying I've got mates looking for the boys [her children] and everything, which can't be classed as breach because he was never served.
>
> (*Sarah, left her relationship, with her young sons*)

Importantly, digital technology can extend the reach and impact of family violence beyond the home.

> I was abused while pregnant and it was a very early stage and I was very sick as well. I was coming to work and I was getting abused at work . . . via social media and text messages and it still stressed me out.
>
> (*Mina, left her relationship, with her one-year-old child*)

The converse of this intrusion through technologies is the denial of access to such technologies and the opportunities they enable for communication (see also George and Harris 2014; McKee 2006). This emerged as a critical issue for several of the women who participated in our study, as described by two women:

> I was isolated without even being aware that that's what was happening. But then it really became clear. . . . He dominated – he had access to my phone, which I wasn't aware of. Everything. Pretty much was hiding messages I was getting from family and friends, which I only found out after I fled.
>
> (*Maria, left her relationship after seven years, with her two daughters*)

> [I didn't know he was] being manipulative because I was a working woman and independent. When I came here I lost my income and being independent goes with it. I can't access – I don't have a phone. I can't access [the

> internet]. If he's – it depends on his mood. If I said to him can I call my [son in South-East Asia?] and then he said it depends on my mood, [you have to wait], you know.
>
> (*Tina, left her relationship after three years*)

Some research has emphasised other technological platforms that facilitate surveillance and abuse. As Levy (2014) observes, 'a huge number of partner "spy" apps exist, with names like *Flexispy, Wife Spy, Girlfriend Spy, Spyera, and ePhoneTracker*' (2014, p. 686). Such apps facilitate constant monitoring and surveillance, providing information about women's movements and activities to perpetrators. Beyond these obviously troubling products, other online interactions and more 'neutral' apps such as GPS, *Find my iPhone*, hidden surveillance cameras and ATM alerts offered by banks can be used as tools to monitor and perpetrate abuse against women. Such abuse is being consistently and increasingly found in contemporary studies of family violence (Burke et al. 2011; Dimond, Fiesler and Bruckman 2011; Mason and Magnet 2011; RCFV 2016; Woodlock 2016). Women's security can be compromised in many different ways by these forms of technological surveillance. Facebook, for example, allows Google Maps to track locations (Mason and Magnet 2011); and other similar common applications may aggregate data from a range of platforms and include geographic location (Dimond, Fiesler and Bruckman 2011). One of our research participants ruefully described the negative financial outcomes of her partner's access to their shared computer and all associated information:

> The only downfall was [the police] were like, 'Well, let him take the computer and the PlayStation because then he'll have something to do and he'll leave you alone'. I'm like yep, fine. I don't care. If he's out of the house, he can take what he wants as long as I have the frigging house. But then he took his name off the mortgage. I can't afford the mortgage so the house – the bank goes, 'Well, we're going to have to repossess'. I go, 'Well *#*, what do I do?' So I was essentially going to be left homeless. I couldn't get a rental because he had destroyed both of our names.
>
> (*Matty, left her relationship after five years, with two children*)

Another outlined the confusion that arose as a result of her ex-partner impersonating her using a mobile phone SIM:

> Okay, so then he got a new SIM and then decided to write up stories pretending that he was her and saying that she was the aggressor and that she was making up all these false accusations.
>
> (*Alia, left her relationship, with five children – speaking through an interpreter*)

The women in our study reported incidents in which intimate photographs were released or in which their release was threatened. This form of abuse, sometimes called 'revenge porn' (see Cavezza and McEwan 2014; Henry and Powell 2015, 2016), is commonly found in other research on family violence.

> He used to actually video-tape her while she was bathing and say, 'Okay, if you do something here is an image of you that I have that I'm going to use against you and that I will put it on the internet'.
>
> (*Alia, left her relationship, with five children – speaking through an interpreter*)

> Now when I found – he used to take photographs of me. Photographs that he took of me doing intimate acts, he showed to other people.
>
> (*Tilly, left her relationship after 10 years, with three children*)

These threats of 'reputational harm' are designed to 'cause fear, distress or alarm' (Cavezza and McEwan 2014, p. 956) and can lead to isolation and insecurity for women.

Woodlock (2016) argues that these technologies operate in multiple ways to facilitate violence against women. These digital platforms can create new forms of private space where men enact violence against women through texting and social media. They can be simultaneously used by perpetrators to strip away women's private spaces, extending the reach of violence, intrusion and surveillance to workplaces, social spaces and elsewhere. Not only do digital technologies extend the possibilities for surveillance, offering ceaseless and extensive access to women's lives, they also have the capacity to isolate women. As Woodlock (2016) identifies, they offer new ways in which perpetrators can become and remain 'omnipresent' in women's lives, even after women have left the relationship. The role of technology-facilitated abuse at the point of separation is also noted by Dimond, Fiesler and Bruckman (2011, p. 419), who argue:, 'Leaving is already difficult for domestic violence survivors, and social networking sites seem to have complicated this issue – they are used for support but also as an additional web of entanglement'.

For this reason, new technologies can exacerbate the risks already present for women at the point of separation – well established as the point of greatest risk of serious and lethal violence to women leaving an abusive male partner (Mahoney 1991). Indeed, recent research has observed that technology-facilitated abuse often escalates during a period of separation (George and Harris 2014; Woodlock 2016).

Resources and responsibility

Alongside recognition of the ways in which technology can be used by a perpetrator to facilitate ongoing abuse, there is acknowledgement of the possibilities of using digital spaces and new technologies to support and empower women who are experiencing family violence (Kee 2006; Mason and Magnet 2011). The most significant opportunity offered has been through the development of online spaces where information about family violence, existing supports, safety plans and many different services have proliferated. These sites are increasingly responsive to women's needs, for example, by offering translation services and online access to response or 'chat'.

The anonymity provided by the internet, where they cannot be identified as they seek information, can enable survivors of family violence to communicate their situation and seek help (Kee 2006), as well as create new avenues to explore and understand gender relations (Kee 2006; Turkle 1995, p. 4). Such spaces are particularly helpful for women who live in remote and rural communities and do not have ready access to support services or who may be unable to leave the family home to seek help.

One aspect of this surge in online information is the development of online services that assist women to identify their experiences and begin to consider options and possibilities to seek change and security for themselves and their children. Recent Australian examples include the *I-Decide* online risk assessment tool and *1800Respect* sites, which allow women to move through a series of questions in order to assess their own sense of safety and level of risk, and offer targeted resources to support them. For example, *I-Decide*, first piloted in 2015, is an online self-assessment tool designed to assess a woman's own perception of her risk and to provide practical and confidential support to victims through an online risk assessment process (Hegarty, Tarzia, Murray, Valpied, Humphreys, Taft, Gold and Glass 2015). The tool is designed for use by women aged between 16 and 50 years and encourages women to rank their goals and aims in seeking help to allow for the referral pathways and support services to be tailored to their individual needs. At the time of writing, *I-Decide* was still progressing through the trial process, and it is unclear to what extent the site will become accessible state-wide. In addition to *I-Decide*, *1800Respect* is a commonwealth government–funded national sexual assault, domestic and family violence telephone and online crisis and trauma counselling service that operates 24 hours daily (RCFV 2016). The service acts as a key referral point by providing information on local support services to callers as well as online resources about recognising abuse, assessing risk, safety planning and support disclosure (RCFV 2016). Both services can be particularly useful for women and children who are seeking help but do not want to engage the criminal justice system.

Another aspect of this move to online information and support has been an increasing recognition by family violence services of the need to protect women online and to empower women to protect themselves online. Online family violence service information and support sites often build in counter-surveillance strategies such as an instant page escape. Classes and resource kits are also offered (see Burke et al. 2011; DVRCV 2015) that assist women to limit perpetrator surveillance opportunities. Women are instructed on how to purchase new SIM cards, create online profiles using aliases and adopt a range of other strategies to address the emerging risks raised by online presence. Information about how to clear 'cookies' and browsing histories often forms part of online resource packages. Such strategies are valuable to ensure the ongoing safety of women leaving an abusive relationship but also support the day-to-day safety of women within abusive relationships, where their partner may be closely monitoring their online behaviours.

Yet as Burke et al. (2011) have observed, the social and contextual framework around such practices has a familiar and potentially troubling aspect. The need for women to 'protect themselves' online reiterates the responsibility put on women to prevent male violence by curtailing aspects of their public and private lives. This construct has the potential to individualise and reprivatise family violence by isolating women and denying them access to the digital spaces that have become an integral part of contemporary public and social life. Dimond Fiesler and Bruckman (2011) and George and Harris (2014) have identified that one way in which women deal with the risks posed by online surveillance is by disappearing from view. As well as changing phones, phone numbers and addresses, women may close down social media accounts and change or delete a range of other online accounts and accesses. While such strategies may reduce the opportunity for perpetrators to spy on or find women, they may also create profound isolation for women and their children. George and Harris (2014) describe women being pressured to 'disappear' from view, intensifying their social isolation, particularly for rural and regional women whose technology-facilitated relationships and interactions may be particularly significant in light of limited opportunities for in-person contact. Woodlock (2016) found that, even when women remove themselves from online platforms, perpetrators use children's social media presence as a vehicle for online harassment of the mother, creating further pressures and fears for those involved. Many of the women interviewed by Burke et al. (2011) gave up social media platforms very reluctantly, as these enabled them to remain active and connected to others. They describe the admission of one of their interviewees, Janelle, that she still checked her sister's Facebook page, even though she knew it was risky. She had moved states, was not in phone contact with family because of the risks to them, and this was the only way she could maintain some form of connection to those closest to her (Burke et al. 2011). Such accounts highlight the irony of the impact of technology-facilitated family violence whereby women leaving an abusive relationship may, through seeking to maintain their own safety, continue to suffer from the isolation and absence of freedom that their abuser sought to impose on them during the relationship. The Royal Commission into Family Violence (2016, volume 1, p. 29) recognised that the impact of such isolation can be particularly heightened for culturally and linguistically diverse women, Aboriginal and Torres Strait Islander women and women with disabilities, when technology-facilitated abuse is used to 'exploit social isolation and language barriers and deliberately isolate culturally and linguistically diverse women further'.

Beyond social isolation for women, there are other impacts of becoming less visible or invisible online. The parallels between the isolation created for women living in an abusive relationship and that experienced by women when they physically flee family violence are clear. If women are required to close down their social communications to stay safe, they rather than the perpetrators are bearing the costs of family violence including isolation, lack of access to necessary or supportive contacts and other potential adverse outcomes such as the inability to apply for rental properties online. Mobile phones may well be the only form of communication available to women if they are fleeing, and the loss of this avenue of communication may result in the loss of access to vital services, social connection or information (Dimond, Fiesler and Bruckman 2011).

In developing processes or discourses that frame women's safety as being secured through silence or absence, there is the potential to create new forms of privacy and individual responsibility in the field of family violence. These strategies reinforce both the opportunity for and responsibility of women to 'act'; and such approaches intermingle with existing discourses that posit family violence as a private or 'domestic' matter, in which confusion about why women do not leave abusive relationships and poor understanding of both the extent of the risks women face and the long-term effects on women and children of sustained violence often continue to make women primarily responsible for their own safety. While women do seek out change and security through processes of disclosure and the search for support, demonstrating great resilience in times of fear, danger and distress, sustained change in women's exposure to family violence will require the acknowledgement of such violence as inherently social and communal, as a shared communal responsibility. It is here that Butler's account of intertwined relationships of violence and vulnerability in *Precarious Life* (2004) offers a critical framework to understand the importance of 'relationality' as a foundational part of women's safety and security rather than as an additional aspect that can be given up. This framework reiterates the need to ensure that responses that support women's safety – from technology-facilitated violence and the associated physical risks – must not reprivatise family violence or individualise the responsibility for confronting and changing that violence.

Holding vulnerability and violence in view

As noted earlier, Butler argues that autonomy is necessary to secure legal rights and entitlements. Moves to change and enhance legal and enforcement responses to family violence have been and continue to be central to achieving change and enhancing security for women. However, Butler proposes that autonomy cannot fully capture the ways in which we are connected and related to each other, and legal entitlements linked to that autonomy, to each of us as bounded individuals, cannot secure what she describes as 'political community'. In particular, Butler (2004, p. 20) argues that we are all necessarily vulnerable to each other, even though such vulnerability creates risk:

> Each of us is constituted politically in part by virtue of the social vulnerability of our bodies. . . . [Such] socially constituted bodies, attached to others, at risk of losing those attachments, exposed to others at risk of violence by virtue of that exposure.

For Butler, dependency and relationality are valuable and important aspects of how we co-exist. We are 'physically dependent on one another, physically vulnerable to one another . . . [this is] another way of imagining community' (Butler 2004, p. 27).

Butler's attention to vulnerability as a necessary aspect of human life has particular resonance in the context of family violence. It is precisely the intimacy of partner relationships that creates the conditions for abuse when perpetrators choose to use violence. It is the intertwined nature of human lives, especially when there are children involved, that extends the spaces and possibilities of perpetrator abuse. Within this schema, new digital technologies can be readily understood as both instruments for and extensions of already-existing connections and threads that bind us together and mean, as Butler says, that no-one is or stays intact.

Butler's purpose in *Precarious Life* is to caution against seeking freedom from violence by trying to separate or disconnect from others. We all live with vulnerability to each other that must be recognised, yet Butler frames this shared vulnerability as critical to creating community. This shared or common vulnerability reinforces our 'collective responsibility' for the physical lives of one another (Butler 2004, p. 44): it is the source of our 'ethical connection' (2004, p. 46) and underpins our relationship and responsibility to each other. 'I cannot think the question of responsibility alone, in isolation from the Other: if I do I have taken myself out of the relational bind that frames the problem of responsibility from the start' (Butler 2004, p. 46).

Butler's creation of a matrix that positions vulnerability and the associated risks of violence as necessarily entwined and as the basis of our 'collective responsibility' resonates with feminist critiques and activism that have urged a rethinking of the nature of family violence and intimate partner violence in particular (Heise, Ellsberg and Gottemoeller 1999; Scutt 1983). Such violence is necessarily understood as communal and relational rather than private or individualised. The conditions for violence against women and children in the context of the family are intimately linked to a continuum of violence and exclusions that constrain and diminish women's social, economic and political opportunities. Although Butler (2004) is clear that neither sex nor gender is fixed, her work has illuminated the schemas of heteronormativity that regulate and define sexed and gendered bodies and create gendered inequalities. Lloyd identifies this operation as 'heteronormative violence . . . violence enacted in the ordering and classification of bodies according to the norms of sex, gender, and sexuality' (2013, p. 825).

In cases of intimate partner violence, this pre-existing regulatory operation that 'defines the kind of behavior attaching to heterosexual masculinity and femininity' (Lloyd 2013, p. 824) is invisible yet constitutive of the social, legal and political interactions and relations that can support or condemn family violence. If we take Butler's proposition about shared vulnerability and collective responsibility seriously, social and political responses to family violence need to be undertaken communally and collectively. Responding to the threats as well as the opportunities offered by new digital technologies needs to be similarly grounded in shared responsibility.

Conclusion

New digital technologies complicate the spaces of family violence in many ways. On the one hand, technology can be used to empower women to seek their own information and safety, but, on the other, technology presents new dangers for women living in or leaving abusive relationships. Social media interactions may work to '[normalize] monitoring practices and data-driven approaches to intimate relations' (Levy 2014, p. 692): the lines identifying technological abuse are blurry and uncertain when constant digital presence and communication are normative. The apps and platforms that support digital communication can readily be turned to monitoring, harassment surveillance and digital stalking, extending the reach and impact of perpetrator presence, meaning that women may feel they are unable to ever escape (Woodlock 2016). George and Harris (2014) argue that the spatial variances of gendered violence need more focused attention if we are to understand the different forms and impact of abuse enabled by digital technologies. Through these technologies, the violence is stretched over time (always) and space (everywhere). The term 'omnipresent' (Woodlock 2016) seems particularly apt to describe such outcomes. The value of online resources for those experiencing family violence cannot be underestimated but must be balanced with recognition that access to such information and the opportunity to act once such information is acquired will often create further danger for women. There is danger and risk both for the individual woman who may be seeking to secure her own safety and for women more generally if responses to technological threats are assigned to women as part of their 'responsibility' to avoid male violence (for example, by closing down Facebook pages or changing contact numbers). Such expectations effectively reprivatise violence against women.

In responding to these new terrains of family violence, it is crucial that the gains that have been made in terms of understanding family violence as a communal and shared responsibility are not lost. Strategies for women to achieve greater safety without having to 'trade off the benefits vs harms' of continuing to use technologies (Dimond, Fiesler and Bruckman 2011) are important. However, resisting any trends that reprivatise family violence, make women responsible for their own online safety and intensify their experiences of isolation or insecurity is critical. As Butler suggests, our shared vulnerabilities make us responsible to and for each other.

Note

1 All authors are from School of Social Sciences, Monash University.

References

Burke, S, Wallen, M, Vail-Smith, K and Knox, D (2011), 'Using technology to control intimate partners: an exploratory study of college undergraduates', *Computers in Human Behavior*, vol. 27, pp. 1162–7.

Butler, J (2004), *Precarious Life: The Powers of Mourning and Violence*, Verso, London and New York.

Cavezza, C and McEwan, T (2014), 'Cyberstalking versus off-line stalking in a forensic sample', *Psychology, Crime & Law*, vol. 20, no. 10, pp. 955–70.

Clarke, R, Wright, P, Balaam, M and McCarthy, J (2013), 'Digital portraits: photo-sharing after domestic violence', *CHI 2013*, April 27–May 2, 2013, Paris, France.

Dimond, J, Fiesler, C and Bruckman, A (2011), 'Domestic violence and information communication technologies', *Interacting With Computers*, vol. 23, pp. 413–21.

DVRCV (2015), *ReCharge: Women's Technology Safety, Legal Resources, Research & Training: National Study Findings*, Legal Service NSW, Domestic Violence Resource Centre Victoria and WESNET.

Garcia-Moreno, C, Jansen, HA, Ellsberg, M, Heise, L and Watts, CH (2006), 'Prevalence of intimate partner violence: findings from the WHO multi-country study on women's health and domestic violence', *The Lancet*, vol. 368, no. 9543, pp. 1260–9.

George, A and Harris, B (2014), *Landscapes of Violence: Women Surviving Family Violence in Regional and Rural Victoria*, Centre for Rural and Regional Law, Deakin University.

Grabosky, P (2016), 'Regulation by ridicule: humorous denigration as a regulatory instrument', *Law, Culture and the Humanities*, vol. 12, pp. 373–400.

Hegarty, K, Tarzia, L, Murray, E, Valpied, J, Humphreys, C, Taft, A, Gold, L and Glass, N (2015), 'Protocol for a randomised controlled trial of a web-based healthy relationship tool and safety decision aid for women experiencing domestic violence (I-DECIDE)', *BMC Public Health*, vol. 15, pp. 736–40.

Heise, L, Ellsberg, M and Gottemoeller, M (1999), *Ending Violence Against Women*, Population Reports December, vol. 27, no. 4, p. 1.

Henry, N and Powell, A (2015), 'Embodied harms gender, shame, and technology-facilitated sexual violence', *Violence Against Women*, vol. 21, pp. 758–79.

Henry, N and Powell, A (2016), 'Technology-facilitated sexual violence: a literature review of empirical research', *Trauma, Violence, & Abuse*, Online First, pp. 1–14, DOI: 10.1177/1524838016650189.

Hoeffler, A and Fearon, J (2015), *Post-2015 Consensus: Conflict and Violence Assessment*, Copenhagen, Copenhagen Consensus Center.

Kee, J (2006), *Cultivating Violence Through Technology? Exploring the Connections between Information Communication Technologies (ICT) and Violence Against Women (VAW)*, Association of Progressive Communications Women's Networking Support Programme.

Levy, K (2014), 'Intimate surveillance', *Idaho Law Revue*, vol. 51, pp. 679–94.

Lloyd, M (2013), 'Heteronormativity and/as violence: the "sexing" of Gwen Arauj', *Hypatia*, vol. 28, no. 4, pp. 818–34.

Mahoney, M (1991), 'Legal images of battered women: redefining the issue of separation', *Michigan Law Review*, vol. 90, no. 1, pp. 1–94.

Mason, C and Magnet, S (2011), 'Surveillance studies and violence against women', *Surveillance & Society*, vol. 10, no. 2, pp. 105–18.

McCulloch, J, Maher, J-M, Fitz-Gibbon, K, Segrave, M and Roffee, J (2016), *Review of the Common Risk Assessment and Management Framework (CRAF): Final Report*, Department of Health and Human Services, Victoria.

Mullen, PE, Pathé, M and Purcell, R (2001), 'Stalking: new constructions of human behaviour', *Australian and New Zealand Journal of Psychiatry*, vol. 35, no. 1, pp. 9–16.

Royal Commission Into Family Violence [RCFV] (2016), *Report and Recommendations*, Royal Commission Into Family Violence, Victoria, available at www.rcfv.com.au/Report-Recommendations (accessed 20 May 2016).

Scutt, J (1983), *Even in the Best of Homes: Violence in the Family*, Sydney, Penguin Books.

Southworth, C, Finn, J, Dawson, S, Fraser, C and Tucker, S (2007), 'Technology, and stalking', *Violence against Women*, vol. 13, no. 8, pp. 842–56.
Turkle, S (1995), *Life on the Screen: Identity in the Age of the Age of Internet*, Touchstone, New York.
Woodlock, D (2015), 'Remote control', *DVRCV Advocate*, Autumn/Winter 2015.
Woodlock, D (2016), 'The abuse of technology in domestic violence and stalking', *Violence against Women*, Online First, pp. 1–19, DOI: 10.1177/1077801216646277.
Yar, M (2013), *Cybercrime and Society*, Sage, London.

2 Gendered exploitation in the digital border crossing? An analysis of the human trafficking and information technology nexus

Sanja Milivojevic and Marie Segrave

> [W]ith the growth of the Internet, human trafficking is now being conducted via telephone and satellite transmissions. Human trafficking is now possible by viewing images of potential girls, bidding [for them] on the Internet, and paying for [services] via the Internet. . . . [I]t is now possible for more human trafficking transaction[s] to occur for purchasing Internet sex slaves.
>
> Joyce Priddy (in Kunze 2010, p. 250)

> We are turning the tables on the traffickers. Just as they are now using technology and the Internet to exploit their victims, we are going to harness technology to stop them.
>
> Barack Obama (Foreign Policy Bulletin 2012)

> We don't have an accurate picture of the scope and nature of [the misuse of technology] and cannot act as effectively as we should. Knowledge is essential for evidence-based policy, and we must fill the information gap.
>
> Yuri Fedotov, Executive Director of United Nations Office on Drugs and Crime (UNODC) (in Latonero 2011, p. v)

Introduction

Borders in the globalised risk society have long ceased to be simply lines in the sand. They are permeable yet enforced, horizontal and vertical, physical and digital. As noted elsewhere, policing the border, now a centrepiece of state intervention, is increasingly mobile, capturing both geographical and virtual borders in the age of globalisation (Milivojevic 2012; Weber 2014). For almost 30 years now, the issue of human trafficking, commonly identified as a cross-border crime par *excellence*, has been dissected, probed and observed by many, arguably with dubious success. While the eradication of human trafficking has been a global priority for quite some time (Bosworth, Hoyle and Madden Dempsey 2011), knowledge about it remains fragmented (see Dragiewicz 2015). As the attention given to the issue in the media, academic and policy contexts continues to grow, the need has never been greater for methodological rigour and 'empirically based rather than ideologically driven' research on human trafficking (Zhang 2009, p. 193).

As trafficking has been firmly located as a cross-border organised crime issue, counter-trafficking policies have predominantly focused on clamping down on trans-border criminal activity and immobilising both offenders and potential/future victims (Segrave and Milivojevic 2015). Technology is emerging as a central concern both in terms of the evolving manifestation of human trafficking, whereby technology is increasingly recognised as a new threat in the facilitation of human trafficking, and as a new battle zone for counter-trafficking efforts (Musto and boyd 2014).

Human trafficking is being reframed via technological developments whereby information technologies make 'many aspects of human trafficking more visible and more traceable, for better and for worse' (boyd et al. 2011, p. 1; see also Musto and boyd 2014; Perer 2012; Sarkar 2015; Vanderschaaf 2013). Indeed, e-trafficking – the term coined to describe human trafficking facilitated/enabled or regulated through the use of the internet and other communication platforms (Milivojevic 2012, p. 73) – is rapidly becoming one of the key concerns of the ever-growing anti-trafficking movement. As Mendel and Sharapov (2016, p. 668) note, 'the use of technology to facilitate human trafficking has attracted increasing attention from anti-trafficking "stakeholders", including governmental and non-governmental sectors over recent years'. In this context, Web 2.0 and social networks have been identified as 'particularly valuable playgrounds' for traffickers (Latonero 2011, p. iv). However, as Latonero (2011, p. iv) points out, 'the extent to which online technologies are used in both sex and labor trafficking is unclear, and the current approach to the question is lacking'. Mendel and Sharapov (2016, p. 668) go a step further, claiming that a range of anti-trafficking actors[1] plays a significant role in 'the construction of ignorance around [the] trafficking [and technology nexus]'.

Importantly, these relations, established in the context of both the trafficking – technology nexus as part of the problem (where technology is identified as a potential facilitator of trafficking) and part of the solution (where technology is perceived as a tool for disrupting trafficking), reinforce a law-and-order response to trafficking, by creating 'new methods of surveillance, exclusion, and expertise' (Musto and boyd 2014, p. 461). We build on Bumiller's examination of the way in which 'society has defined sexual violence as a social problem and how this creates policies that reinforce stereotypical assumptions about women's dependency and the character of intimate violence' (2008, p. xiv). We extend this argument to consider how 'the large-scale expansion of legal and government efforts to counteract the threat of sexual violence [that] has transformed the everyday relationships between the state and women as both actual and potential victims' is pertinent to the examination of responses to e-trafficking (2008, p. xiv). As we develop later, what can be seen in the emerging articulation of the threat of e-trafficking and the logic underpinning the required response is a prevailing focus on women as potential victims. We argue, as Musto and boyd (2014, p. 4) have, that 'technologically mediated anti-trafficking interventions appear to blur the boundaries between sex trafficking and sex work and subject individuals deemed "at-risk" to new forms of surveillance'. While, as Musto and boyd go on to assert, this 'risks perpetuating harms against the people they aim to assist' by making victims of forced labour and voluntary sex workers at increased vulnerability of 'heightened law enforcement surveillance and carceral oversight' (2014, pp. 4–5), we argue that it also silences any consideration of the limitations and impacts of state policies, particularly migration policies, and the refusal to recognise women as active migrants utilising their agency to seek opportunities to better their lives. To build this argument, we begin by mapping the development of the trafficking–technology nexus.

Human trafficking and information technologies: a precarious connection or an elusive fantasy?

The claim that human trafficking is innately linked to mobile and information technologies is anything but new. The initial (albeit at first sporadic) voices warned about the danger of the internet and its potential role in facilitating trafficking nearly 20 years ago. One of the pioneers in exploring the connection between trafficking and information technologies was Donna Hughes, an American abolitionist[2] and prolific scholar on the topic of sex trafficking. In the late 1990s, Hughes warned that 'the sex industry and the Internet have become partners in the global exploitation of women and children' (Hughes 2000, p. 35). She also suggested that '[u]sing new technologies, sexual predators and pimps stalk women and children. . . . These technologies enable sexual predators to harm and exploit women and children efficiently and, [sic] anonymously' (Hughes 2002, p. 129). Alarmingly, Hughes also argued that '[t]he sexual exploitation of women and children is a global human rights crisis that is being escalated by the use of new technologies' (Hughes 2002, p. 129) and that '[t]he use of new communication and information technologies for the sexual exploitation of women and children continues to grow with the increased number of users on the Internet' (Hughes 2002, p. 148). At that early stage of the awareness raising around human trafficking and technology, a significant component of the international anti-trafficking architecture was founded upon the conflation of women and children as equal in the risk narrative, necessarily infantilising women and reducing them to passive victims. Importantly, as we will elaborate on later, most of these claims were based on limited, if any, academic research. While more nuanced accounts have emerged (see, for example, Doezema 2010; Dragiewicz 2015) in the e-trafficking context, this familiar trope has arisen again within the arguments surrounding the threat that lurks online – on the Dark Web,[3] social media or online advertising. The technology itself has been constructed as a new (and extreme) form of violence, so much so that sweeping and somewhat apocalyptic claims like Hughes's have prompted supra-national international organisations and state governments into action.

On the strength of these arguments and her body of work, Donna Hughes was appointed to lead the Council of Europe (COE) 'Group of Specialists on the Impact of the Use of New Technologies on Trafficking in Human Beings for the Purpose of Sexual Exploitation' (hereafter, the Group), which produced three key reports on the topic in the early 2000s. The first two reports were authored by Hughes (2001a, 2001b), and in this capacity she revitalised the focus on women regarded to be 'at risk': she argued, for example, that there is an 'epidemic of trafficking for sexual exploitation on the web' (Hughes 2001a, p. 24). The final report released by the Group concluded that virtual trafficking is a negative manifestation of globalisation. It stated that 'the impact of the use of new information technologies on trafficking in human beings for the purpose of sexual exploitation is now seen as a serious issue' (COE 2003, p. 15) that consequently requires the immediate response of the international community and policy makers. Evident in these reports was a focus on sex trafficking as opposed to human trafficking in its various manifestations. In this context, technology was constructed as a new vehicle through which the threat of victimisation was materialised. Underpinning this focus was the gendered politics of women as at risk of victimisation: the arguments were presented as objective 'fact', despite being developed and informed by a particular viewpoint regarding violence against women and women's agency, specifically in the area of prostitution/sex work. As these ideas have increasingly captured the international imagination, what we have witnessed in both the production of stories of trafficking victimisation and policy and legislation around this issue is the legitimising of interventions that pay little heed to the impact of women from the Global South being deemed 'undesirable' citizens (see Andrijasevic and Mai 2016, p. 1). Indeed, women and children as victims have been placed at the heart of the technology–trafficking nexus, with the consequence of consolidating 'dominant expectations that certain sexual behaviours are extraordinarily risky' (Musto and boyd 2014, p. 6).

By 2008, the UN Global Initiative to Fight Human Trafficking (UN.GIFT) included e-trafficking in its fifth session of the Conference of the Parties to the UN Convention on Transnational Organised Crime, stating that 'technology offers traffickers more creative and complex ways to commit their crimes while at the same time, presenting the global anti-trafficking community with more opportunities to respond to it' (UN.GIFT 2008, p. 2). The United States (US) *Trafficking in Persons Report* (hereafter, the TIP report) first mentioned the links between trafficking and information technologies in 2011 (US Department of State [USDOS] 2011), while the focus on the trafficking–technology nexus in Europe flourished during 2013 and 2014. In 2013, the UK's Centre for Social Justice released a report titled *It Happens Here: Equipping the United Kingdom to Fight Modern Slavery*, which noted 'the role of the Internet in recruiting and "selling" people into "modern slavery"' (Mendel and Sharapov 2016, p. 668). Another report by the same organisation followed in 2015 (*Modern Response to Modern Slavery*), warning of 'the growing speed with which organised criminal groups use cyber techniques' to enslave victims (Mendel and Sharapov 2016, p. 669). EUROPOL jumped on board in 2014, noting that '[t]he link between the use of the Internet, cybercrime and [trafficking in human beings] activities is now stronger than ever before, as human trafficking networks increasingly exploit "online market places" to their benefit, for example in the recruitment of victims' (EUROPOL 2014). The evidence to support these claims was spurious at best. Yet despite the absence of rigorous, empirical data, the assertion that the digital sphere is a new battlefront in combating trafficking has informed a range of technological solutions to the problem of trafficking.

The internet and social media have long been identified as potentially helpful tools for educating people (both the public and potential victims) about human trafficking, as well as platforms for the digital activism of anti-trafficking activists (Gong 2015). In 2011, the TIP report stated that '[n]ew media may emerge as a powerful tool for identifying victims and bringing their traffickers to justice' (USDOS 2011, p. 35). Similarly, the 2012 *State of Human Trafficking* report argued that, 'while technology is being used to perpetrate human trafficking, the same technology can provide a digital trail' that 'offers greater potential opportunity for tracking traffickers' and johns' communications, movements, and transactions' (California Department of Justice 2012, pp. 7, 65). Various software and predictive analytics have been developed to 'detect possible cases of online sex trafficking activity' (Latonero 2011, p. v). Google, Microsoft and other tech giants have been spending millions of dollars on initiatives to combat trafficking (Latonero 2012, p. 17). The Dark Web is penetrated by new programs with advanced internet search capabilities (Greenmeier 2015), while cloud-based analytics such as DIG (domain-specific insight graphs) aim to automatically trawl the internet for escort ads (*Wired*, 19 May 2015). Thus, the trafficking–technology nexus has been firmly established in both the 'part of the problem' and 'part of the solution' contexts. It is now necessary to look closely at how this trafficking–technology conundrum has rapidly converted into policy. It is also important to consider the role of gender in shaping both the assumptions about victims and the design of counter-trafficking strategies online, as we do next.

Dissecting the trafficking–technology nexus: from fantasy to policy

O'Malley (2008) reminds us that in today's risk society, managing risk is the central tenet of governance, while victims of crime play a pivotal role in risk-centred crime control strategies. This is particularly accurate when it comes to e-crimes. The nature of information technologies and the internet – their perceived unregulated and deterritorialised nature, and ultimately ungovernability – has been singled out as one of the key reasons why responses to technology-enabled crimes (including trafficking) need to be uncompromising and aligned with a law-and-order framework, already deployed to combat terrestrial crimes. The internet's global reach and unprecedented growth and the potential global impact of crimes committed with the assistance of or via online platforms have arguably exacerbated the already heightened vulnerability of the Global North (Milivojevic 2012). Various threats that lurk on the internet, from cyberterrorism to online child sexual abuse, are undoubtedly at the very pinnacle of newsworthiness in the twenty-first century (see, for example, Wall 2013).

The threats that lurk online are often perceived to be clandestine, covert, hidden and deeply embedded in the anonymous World Wide Web (see COE 2003,

p. 93; Farley, Franzblau and Kennedy 2013; Vitale 2012). The trafficking 'other' (the offender) supposedly thrives in such an environment. As Mendel and Sharapov (2014, p. 3) note, 'the increasing accessibility and developing technologies of the Internet and digital networks are enabling traffickers to operate with increased efficiency and anonymity' (see also Vanderschaaf 2013). E-trafficking thus becomes a new form of violence and, in the context of trafficking, is seen as violence that disproportionately impacts on women and children and is almost exclusively linked to exploitation in the sex industry. Further, the difficulty of prosecuting e-crimes and the absence of criminal accountability for those who engage in cybercrimes, including e-trafficking, are articulated as the justification for the significant surveillance of online activities and the internet more broadly. As described earlier, strategies to identify and alert to suspicious behaviour and sites are largely based on surveillance and trawling the internet, with the ultimate goal of protecting the vulnerable – women and children. This echoes Bumiller's argument that, in place of recognising what limits women's agency and ability to make decisions for themselves and/or to act on those decisions, surveillance of women is increased for their protection. In the context of human trafficking, surveillance and online 'education' are often therefore gendered bordering practices – enabling the efforts to actively prevent women perceived as undesirable migrant 'others', particularly in the Global North, from seeking opportunities to migrate, purportedly for their 'protection' (see Segrave, Milivojevic and Pickering 2009).

As we argue later in the chapter, given that trafficking is located at the international level within the *Convention against Transnational Organised Crime*, various state and non-state agencies, media, faith-based organisations and feminists call for more rigorous state intervention, tougher legislation, surveillance of online activities of traffickers and potential victims and unconditional cooperation with law enforcement in investigating these crimes. The law-and-order repertoire of state coercive powers is indeed perceived as a key weapon in the 'war on trafficking' (Milivojevic 2012), which creates a challenging platform upon which to advocate for protections online, such as anonymity of data usage and/or providing safe platforms for women who actively seek opportunities online to migrate regularly and irregularly. The logic that posits that technology is part of the solution to the trafficking 'problem', in which the key aim is to identify, apprehend and prosecute traffickers, is flawed. There is also limited, if any, understanding of how technology is used to facilitate women's (and men's) migration strategies.

Populist analogies highlight the ease with which technologies can be misused (COE 2003, p. 93). For example, Andrea Powell has explained that ordering sex with a minor online is 'as easy as ordering pizza' (Andrea Powell, anti-trafficking group FAIR Fund, cited in Thakor and boyd 2013, p. 279), serving the dual purpose of spreading panic about technological advances that enable the intrinsically evil practices of child sex abuse and prostitution to flourish while yet again equating women and children as equally passive and vulnerable. The already familiar threat of 'online sex predators' furthers the notion that the internet and its advances are contributing to the growth of the sex industry and thus, implicitly, trafficking itself. As Farley, Franzblau and Kennedy (2013, p. 1043) note, '[t]he Internet and computer technology have been developed and exploited by sex businesses to offer prostitution to men across the globe'. One of the most extreme examples of this approach is the report by Shared Hope International on the topic, which concluded that '[t]echnology has become the single greatest facilitator of the commercial sex trade' and thus '[a]s one of the most technologically advanced countries in the world, the U.S. faces the challenges of combating facilitation of sex tourism and sex trafficking markets by technology' (cited in Latonero 2011, p. 14). These arguments echo early arguments in the counter-trafficking space, where human trafficking was equated to sex trafficking, as the real (and almost only) focus was on sex work or, for many of those advocating in this area, prostitution.[4] For some women, pursuing avenues to migrate to another country requires complex negotiations with a range of actors; the online space creates more opportunities to do so (see Piper and Roces 2003 on the complexities of women seeking migration and marriage opportunities online). Yet again, the gendered politics of the decades-long abolitionist agenda becomes codified in counter-trafficking strategies, where the threat is sex work and not, for example, women's limited ability to migrate and the ways in which limitations on their mobility produce and sustain their vulnerability (see Segrave, Milivojevic and Pickering 2009). Thus, while some construct technology as a potential advancement for sex work/the sex industry, technology as such is translated into a threat in its own right.

Even when the technology itself is not seen as criminogenic or the factor that increases trafficking, it is argued that technology 'has made the activities [of traffickers] easier and introduced new ways of trafficking in human beings for the purpose of sexual exploitation' (COE 2003, p. 93, emphasis in original). As Latonero (2011, p. iv) notes, '[w]hile traditional channels of trafficking remain in place, online technologies give traffickers the unprecedented ability to exploit a greater number of victims and advertise their services across geographic boundaries'. It is noteworthy that Latonero appears, like others, to be conflating smuggling and human trafficking. Also importantly, how do we know that all those promoting their service to assist someone to migrate with a promise of labour at the point of destination are traffickers or that their efforts will result in trafficking? Indeed, and yet again, assumptions are made about the technology–trafficking nexus that produce 'certain types of ignorance' (Mendel and Sharapov 2016, p. 666) and simultaneously silence more nuanced understandings of technology as a mobility tool in migration processes. The reality is that we do not know to what extent and how this nexus really works, and therefore rigorous research in this area is critical. Perpetuating the idea that the internet enables human trafficking to transgress geographical borders creates an illusion of an omnipotent threat that is primarily posed to people (women) in the Global South whose access to migration opportunities are limited. Yet in framing trafficking as a technological problem, anti-trafficking actors have successfully combined anti-prostitution, anti-immigration and the law-and-order agenda while at the same time disregarding (mostly women's) agency.

Most of these interventions have been directly or indirectly underpinned by research on the topic conducted throughout the 2000s and beyond. As we argued elsewhere, most of the research (as well as mainstream media reporting) on the topic has blurred the boundaries between the sex industry and sex trafficking (Milivojevic 2012). As Donna Hughes (2000, p. 41) puts it:

> [f]ew women choose to be in the sex industry. If any choice is involved it is usually the last choice, when no other options are available. Each year hundreds of Ukrainian women are trafficked into the United States. Most of the women believe they will be employed as waitresses or nannies, only to be forced into the sex industry using debt bondage once they arrive.

Farley, Franzblau and Kennedy (2013, pp. 1039–40) take this argument further, claiming that '[f]alse distinctions have been erected between online and offline prostitution, child and adult prostitution, indoor and outdoor prostitution, pornography and prostitution, legal and illegal prostitution, and prostitution and trafficking'.

Most importantly, however, some of the findings of e-trafficking reports have been based on highly speculative accounts, as seen in the examples provided here:

> [T]here is little documentation on the use of new information technologies for criminal purposes by traffickers and pimps, but *there is no reason to assume* they are not using the latest technologies for their transnational or local activities. . . . It is *likely* [that they] are using new technologies for ease of communication and to avoid detection. . . . As more cases of trafficking for the purpose of sexual exploitation are uncovered, the details of their operations will *most likely* reveal an increased use of electronic communications.
>
> (COE 2003, p. 23, emphasis added)

> Although there is little information on the use of the new information technologies for sexual exploitation purposes by traffickers, *there is no reason to doubt* that they are using the latest technologies for trafficking purposes too.
>
> (Sarkar 2015, p. 59, emphasis added; for more examples, see Milivojevic 2012)

It is also clear that what is occurring is a blurring of divisions across human trafficking, child abuse and forced prostitution. Yet such arguments silence women's experiences. In the absence of evidence, there is a strong, gendered and moralising argument. This is apparent in many reports on e-trafficking: for example, the COE reported 'strong indications' (COE 2003, p. 48) for e-trafficking – a particularly meaningless empirical or quantitative assessment. Much like research on its terrestrial counterpart, the research related to online trafficking facilitation often violates the 'standard canons of the social science inquiry' (Weitzer 2012, p. 1339). Peer-reviewed journals have frequently published non-empirical research on trafficking (Gozdziak 2014), making the knowledge base on the topic rather skewed. Yet once made, these broad claims about the extent and nature of trafficking 'have been perpetuated and remain virtually unchallenged in contemporary academic and populist debate' (Milivojevic 2012, p. 80). Alarmingly, the scholarship based on non-empirical research and/or limited evidence has been very effective in informing policy[5] (see Latonero 2011; Mendel and Sharapov 2016; Musto and boyd 2014).

Prompted by the research projects outlined earlier, the COE in 2005 extended the definition of trafficking to include the use of information technologies in the recruitment of victims (Article 79 of the *Convention on Action against Trafficking in Human Beings* [CETS No. 197]; also in Sykiotou 2007, p. 8). The *EU Strategy towards the Eradication of Trafficking in Human Beings 2012–2016* identified the internet as an emerging platform for the recruitment of victims (Mendel and Sharapov 2016, p. 671). In the UK, the *Modern Slavery Act* and the *Modern Slavery Strategy* passed in March 2015 were largely influenced by the Centre for Social Justice research that favoured tabloid media reporting on the trafficking–technology nexus rather than academic research (Mendel and Sharapov 2016, p. 668). Offering no evidence for such claims, the strategy argued that '[i]nternet infrastructure [is] used by serious and organised criminal groups' to recruit victims (Mendel and Sharapov 2016, p. 669; see also Farley, Franzblau and Kennedy 2013). In the US, as Musto and boyd (2014, p. 462) point out, recent legislative changes based on expectations that information technologies can disrupt trafficking proposed public–private partnerships in addressing the issue; heightened collaboration between states, non-government organisations (NGOs) and corporations; legislative changes; and increased surveillance and monitoring of the cyberspace and mobile networks as the most effective strategies to combat trafficking (see also COE 2003). Importantly, these policy changes have had many (albeit perhaps collateral) adverse implications for (actual or potential) victims of trafficking.

The implications: from preventing harm to creating harm

Latonero (2011, p. v) notes that '[t]he ultimate beneficiaries of any technological intervention should be the victims and survivors of human trafficking'. However, he warns that '[w]hile the benefits [of such intervention] may appear clear at first, experts should be consulted to evaluate whether tools have the potential to cause inadvertent harm'. Indeed, as Musto and boyd (2014, p. 463) note, the 3Rs (rescue, rehabilitation, reintegration) and 4Ps (prevention, protection, prosecution, partnerships) framework underpinning terrestrial anti-trafficking interventions have transformed into the 4As framework of e-trafficking (*awareness* of internet-based websites and locations promoting trafficking; *amassment* of data collected by law enforcement to locate and prosecute traffickers and rescue victims; *augmentation* of traditional surveillance techniques; and *automated/algorithmic* techniques increasingly used to identify and disrupt trafficking networks). Ultimately, the beneficiary of such an approach is the security state, 'pre-occupied with the management of the marginalized and socially excluded' (Hallsworth and Lea 2011, p. 141). E-trafficking's powerful offender (an organised crime trafficker lurking in the darkest corners of the deep web) and innocent victim (a naïve and innocent girl from the Global South who responds to ads on the internet) require an uncompromising response. The state's punitive apparatus is expanding, its surveillance omnipresent, and every participant in the anti-trafficking movement is called to the duty – from NGO activists to internet users and potential victims. This development is not new: we witnessed the rise of unlikely coalitions in the terrestrial anti-trafficking movement, with radical feminists, right-wing governments and faith-based organisations joining forces to rescue innocent victims of the sex industry (Milivojevic and Pickering 2008). Yet in the case of e-trafficking, we can identify new harms alongside the already known harms brought by terrestrial anti-trafficking interventions. These harms are created through the expansion of police powers and the limitation of privacy through surveillance, data retention and disruption of technology; the further limitation of the mobility of mobile bodies from the Global South; and over-criminalisation and the new punitiveness in the context of e-trafficking. These harms are necessarily gendered, in part because women are seen as most at risk or vulnerable and are thus the targets, which necessarily means that the millions of men who identify and travel for work opportunities in other nations, only to find themselves forced into labour for little or no pay and threatened with deportation, are also ignored. We identify three key consequences of such an approach.

Surveillance, data retention and disabling technology

Katja Franko Aas (2011) reminds us that existing strategies for the surveillance of mobility in both the terrestrial and digital worlds in the context of cross-border crimes are considerable and growing. The trafficking–technology nexus further increases the ongoing surveillance of youth and adults in the sex industry and, we would argue, in the digital spaces more broadly; yet there is limited understanding of the counterproductive and potentially harmful impacts of this increased surveillance (Mendel and Sharapov 2016; Musto and boyd 2014). We know also that many of these practices tend to reinforce the notion that women and children are most 'at risk', resulting yet again in the assumption that men are not victims of forced labour or human trafficking and that men can negotiate online spaces with greater acumen than women.

The rescue and rehabilitation narrative that underpins such interventions results in restrictions being imposed on the lives of the objects of surveillance. Such surveillance powers are mostly proposed and assigned to law enforcement agencies, both national and international (COE 2003, p. 97). The pre-emptive approach to 'preventing' trafficking results in expanding the capacity of search engines to look for 'suspicious' content online, ever-increasing requests for data retention as a prerequisite to combatting e-trafficking and over-regulation of the content on the internet, which often includes censoring unwanted content (that is, shutting down sex industry–related websites). A stark example of this approach is the case of Craigslist, a free online classified site on which the Adult Services section came under scrutiny in the media as an 'increasing number of news outlets reported on traffickers using Craigslist postings to advertise trafficked persons' (Latonero

2011, p. 13).[6] The requests for the whole internet to be regulated through national and international regulatory mechanisms in order to prevent future victimisation through trafficking are many and growing (see, for example, Vitale 2012). Yet assumptions that technology plays an important role in facilitating human trafficking in this and many other cases have been supported by shoddy research that lacks methodological rigour[7] and definitional clarity. As Thakor and boyd (2013, p. 278) note, the data used to 'prove' that shutting down Craigslist was effective 'made no distinction between sex work and sex trafficking'. The continuation of this blurring is doing significant damage to any effort to change women's and men's vulnerability to exploitation related to migration and their migration status.

Increased immobility

Decades of research in terrestrial trafficking led some researchers to believe that when 'states enact law and policy or sign conventions to stop trafficking, they overwhelmingly opt to intensify border and immigration control' (Vance 2011, p. 935). It is well documented that anti-trafficking campaigns often result in limiting the mobility of mobile bodies from the Global South (Milivojevic and Pickering 2008; Nieuwenhuys and Pecoud 2007). It is also increasingly well documented that gendered bordering is ever present, highlighted in the decisions made at the border, such as judging women according to their clothes or the items found in their suitcases (see Pickering and Ham 2014). As border migration controls expand to new frontiers, away from physical, geographical borders, new strategies of control and immobilisation have been deployed in both the terrestrial and digital worlds. In the trafficking context, potential victimisation has often been used as (an often effective) justification for immobilisation (see, for example, Milivojevic and Pickering 2008). As the trafficking–technology nexus further accelerates the perceived threat of trafficking, with 'grossly exaggerated numbers' (Vance 2011, p. 935) of offenders and victims online serving the purpose to further extend existing laws and to pass new policies that ultimately limit mobility of mobile bodies from the Global South. Defining information technology as a dangerous platform where potential victimisation is only a matter of time aims to immobilise mobile bodies in the country of destination before they embark on a precarious journey towards the Global North (see Mendel and Sharapov 2016 for a discussion of the Ukraine context).

Over-criminalisation and the new punitiveness

Emerging from the trafficking-as-organised-crime narrative, which posits trafficking as even more sinister when committed online, are calls for increased sanctions for traffickers, as well as for the criminal liability of creators of abusive websites and hyperlinks on the internet (along with a range of other interventions required by nation-states on the national and international level – see COE 2003, pp. 95–6). Such efforts, producing reports of threats and perpetuating perception of women as always at risk, frame this issue as one of victimisation, whereby criminalisation and surveillance represent the antidote. As Andrijasevic and Mai (2016, p. 5) argue, such research mobilises:

> stereotypical narratives . . . about sexuality, gender, class and race that end up by demarcating people's entitlement to social mobility and citizenship in increasingly unequal times. By focusing on the spectacular and criminal exploitation of a minority of victims, they legitimise restrictive migration policies and anti-trafficking interventions containing racially and socially 'undesirable' groups and exacerbating migrant workers' exposure to trafficking. . . . [This] conveniently distract[s] the global public from their increasing and shared day-to-day exploitability as workers because of the systematic erosion of labour rights globally. In doing so, they become complicit in the perpetuation of the very social inequalities, hierarchies and conflicts that allow exploitation and trafficking to occur.

Focusing on decreasing the ability of people to utilise the internet and to explore opportunities without fear of constant surveillance serves ultimately as a distraction from the real issues at play. It also stifles the ability to challenge the utilisation of surveillance measures online, as such challenges are opposed by the assertion of the need to protect vulnerable women and children. This is an extension of Bumiller's argument of the co-option of strategies addressing violence against women and yet another example of the harm generated by purported protective measures (see Dagistanli and Milivojevic 2013; Kapur 2002).

Conclusion

A growing scholarship in social sciences and criminology is scrutinising the appropriation of the women's rights agenda in response to both domestic and cross-border crimes in order to promote conservative law-and-order and exclusionary policies (Dagistanli and Milivojevic 2013). Importantly, moral outrage about (prospective) violated women's bodies is equally distributed in both the terrestrial and digital worlds. A robust mythology so deeply embedded in our understanding of terrestrial trafficking, where claims about the practice are often unsubstantiated (Weitzer 2012, p. 1337), has been transferred to the illusory construct of e-trafficking. Underpinned by a political and moral agenda and values rather than robust research and data, this approach frames technology as a threat and ultimately as a new form of violence against women. Bold and sweeping statements about the trafficking–technology nexus, often unsubstantiated and unquestioned, spread across the anti-trafficking movement and penetrate the media and public spaces. In the context of terrestrial trafficking, we have seen how 'quick and dirty' research (Kelly 2002, p. 7) has been rapidly translated into questionable policy. A predominant e-trafficking agenda is yet another example of this approach.

The outcome of such an approach is a range of supposedly protective measures that result in further securitisation of digital spaces, simplification of the trafficking debate, and extreme responses that alienate, immobilise and amplify harm. The threat of e-trafficking, amalgamated with the gendered narrative of naïve women who cannot navigate online spaces, has been addressed through the rescue narrative and a range of protectionist interventions that impede women's agency. As such, e-trafficking provides yet another example of how a protectionist agenda additionally limits women's human rights. We argue that what is evident here is the advancing of Bumiller's argument surrounding the co-option of the feminist agenda. However, in the area of human trafficking, this agenda was shaped by radical, abolitionist feminist legacy that focuses on women as victims and barely strays from the decades-old debate around prostitution as sexual violence versus labour. This narrative lacks nuance and connection with women's experiences and absents the impact of policy from the purview of those designing and implementing counter-trafficking strategies. The damage done by continually referring to women and children in the same breath is to simplify the conditions that give rise to women's exploitation and, at the same time, to infantilise women such that they are no longer migrant labourers who actively – and largely safely – participate in both the terrestrial and online worlds but naïve, desperate, uneducated prey. Such an approach has failed miserably to address terrestrial trafficking; it is time to stop repeating the same mistakes.

Bumiller writes that the codification of violence against women as a form of human rights abuse 'has important implications . . . in relation to the ways it potentially contributes to the growing criminalisation of sexual violence that, in turn, expands the policing functions of the modern state' (2008, p. 134). We argue, building on Bumiller, that what happens in these overt efforts to 'protect women' from themselves (or their own naivety) is that the 'modern state becomes an oppressor itself' (2008, p. 138). It does so by continuing to limit women's agency through increasing the surveillance of their activities online, in turn further decreasing women's mobility. We have argued elsewhere that the consequence of relying on the traditional criminalisation tropes of victim and offender and the role of law is that we do not question the broader ways in which the politics and power of the state result in women being placed in situations in which their only migration pathway is irregular (see Segrave and Milivojevic 2015; Segrave, Milivojevic and Pickering 2009).

Importantly, the logic of technology constructed in the e-trafficking context is problematic and flawed, as technology is framed as 'part of the problem' or 'part of the solution'. As such, the contemporary technology–trafficking rhetoric blurs many distinctions when it comes to mobility and victimisation in the digital age. A new approach needs to be resistant to simplified, politicised and moralised narratives. We need to resist the notion that technology is intrinsically problematic and risky. We argue that what is known about the connection between human trafficking broadly (that is, beyond sex trafficking) and technology requires more careful, robust research. So, too, we argue that such research has to ensure that the broader context of policies and practices that inhibit women in multiple ways – from migration controls to limitations on labour protections – must be recognised in any analysis of the impact and effectiveness of counter-trafficking solutions online.

Notes

1 Such as the UK's Centre for Social Justice and the European Cybercrime Centre.
2 Abolitionists campaign for the recognition of prostitution as a form of violence against women and against the recognition of sex work as a valid form of labour.
3 The Dark Web is the World Wide Web content that exists on darknet networks. The content is not indexed and can be accessed only through specific software, configurations or password-authorised access. The Dark Web is often associated with criminal activity (see, for example, Romeo 2016).
4 Abolitionist scholars and radical feminist do not use the term 'sex work', as they want to emphasise that prostitution is a 'quintessential form of male domination over women – the epitome of women's subordination, degradation, and victimization' (Weitzer 2005, p. 211).
5 An infamous example is the study of the online child sex industry in the US in which the method applied to determine the number of juvenile prostitutes was counting pictures of young-looking women online. The study was nevertheless introduced to the congressional testimony addressing the issue of minor sex trafficking in the US (Latonero 2011, p. 15).
6 In September 2010, Craigslist shut down its Adult Services service, 'following years of being dogged on the issue by states attorney general and some human trafficking groups, which eventually led to a hearing before Congress' (Siegel 2010; see also Thakor and boyd 2013).
7 Often, these reports rely on media reporting on the issue, taking it at face value and not engaging critically with the content reported by media outlets, or are based on a limited number of qualitative interviews (see, for example, Farley, Franzblau and Kennedy 2013; Hughes 2002; Kunze 2010 on the use of social networking sites for the recruitment of victims). For more on this, see Milivojevic (2012).

References

Aas, K (2011), '"Crimmigrant" bodies and bona fide travellers: surveillance, citizenship and global governance', *Theoretical Criminology*, vol. 15, no. 3, pp. 331–46.

Andrijasevic, R and Mai, N (2016), 'Editorial: trafficking (in) representations – understanding the recurring appeal of victimhood and slavery in neoliberal times', *Anti-Trafficking Review*, vol. 7, pp. 1–10.

Bosworth, M, Hoyle, C and Madden Dempsey, M (2011), 'Researching trafficked women: on institutional resistance and the limits to feminist reflexivity', *Qualitative Inquiry*, vol. 17, no. 9, pp. 769–79.

boyd, d, Casteel, H, Thakor, M and Johnson, R (2011), 'Human trafficking and technology: a framework for understanding the role of technology in the commercial sexual exploitation of children in the U.S.', Microsoft research, available at http://research.microsoft.com/en-us/collaboration/focus/education/htframework-2011.pdf (accessed 23 June 2016).

Bumiller, K (2008), *In an Abusive State: How Neoliberalism Appropriated the Feminist Movement Against Sexual Violence*, Duke University Press, Durham.

California Department of Justice (2012), *The State of Human Trafficking in California*, available at http://oag.ca.gov/human-trafficking (accessed 21 June 2016).

Council of Europe [COE] (2003), *Group of Specialists on the Impact of the Use of New Information Technologies on Trafficking in Human Beings for the Purpose of Sexual Exploitation*, Strasbourg, available at https://ec.europa.eu/anti-trafficking/publications/group-specialists-impact-use-new-information-technologies-trafficking-human-beings_en (accessed 24 June 2016).

Dagistanli, S and Milivojevic, S (2013), 'Appropriating the rights of women: moral panics, victims and exclusionary agendas in domestic and cross-border crimes', *Women's Studies International Forum*, vol. 40, pp. 230–42.

Doezema, J (2010), *Sex Slaves and Discourse Masters: The Construction of Trafficking*, Zed Books, London and New York.

Dragiewicz, M (ed.) (2015), *Global Human Trafficking: Critical Issues and Contexts*, Routledge, Oxon.

EUROPOL (2014), *Increased Focus on the Link Between the Internet and Human Trafficking*, available at www.europol.europa.eu/content/increased-focus-link-between-internet-and-human-trafficking (accessed 24 October 2016).

Farley, M, Franzblau, K and Kennedy, A (2013), 'Online prostitution and trafficking', *Albany Law Review*, vol. 77, no. 3, pp. 1039–67.

Foreign Policy Bulletin (2012), 'Obama pledges to use technology to target human trafficking, simplify visas for trafficking victims', *Foreign Policy Bulletin – The Documentary Record of United States Foreign Policy*, vol. 22, no. 4, pp. 26–45.

Gong, R (2015), 'Indignation, inspiration, and interaction on the internet: emotion work online in the anti-human trafficking movement', *Journal of Technology in Human Services*, vol. 33, no. 1, pp. 87–103.

Gozdziak, E (2014), 'Data matters: issues and challenges for research on trafficking', in M Dragiewicz (ed.), *Global Human Trafficking*, Taylor & Francis, London, pp. 23–38.

Greenmeier, L (2015), 'Human traffickers caught on hidden internet', *Scientific American*, available at www.scientificamerican.com/article/human-traffickers-caught-on-hidden-internet/ (accessed 23 June 2016).

Hallsworth, S and Lea, J (2011), 'Reconstructing leviathan: emergency contours of the security state', *Theoretical Criminology*, vol. 15, no. 2, pp. 141–57.

Hughes, D (2000), 'The internet and sex industries: partners in global sexual exploitation', *IEEE Technology and Society Magazine*, Spring 2000.

Hughes, D (2001a), *The Impact of the Use of New Communications and Information Technologies on Trafficking in Human Beings for Sexual Exploitation: A Study of Users*, Council of Europe, available at http://uri.academia.edu/DonnaHughes (accessed 23 June 2016).

Hughes, D (2001b), *The Impact of the Use of New Communications and Information Technologies on Trafficking in Human Beings for Sexual Exploitation: Marriage Agencies and Images*, Council of Europe, available at http://uri.academia.edu/DonnaHughes (accessed 23 June 2016).

Hughes, D (2002), 'The use of new communications and information technologies for sexual exploitation of women and children', *Hastings Women's Law Journal*, vol. 13, no. 1, pp. 129–48.

Kapur, R (2002), 'The tragedy of victimization rhetoric: resurrecting the "native" subject in international/post-colonial feminist legal politics', *Harvard Human Rights Journal*, vol. 15, no. 1, pp. 1–38.

Kelly, L (2002), *Journeys of Jeopardy: A Review of Research on Trafficking in Women and Children in Europe*, International Organization for Migration, Geneva.

Kunze, E (2010), 'Sex trafficking via the internet: how international agreements address the problem and fail to go far enough', *Journal of High Technology Law*, vol. 10, no. 2, pp. 241–90.

Latonero, M (2011), *Human Trafficking Online: The Role of Social Networking Sites and Online Classifieds*, University of Southern California, available at http://technologyandtrafficking.usc.edu/files/2011/09/HumanTrafficking_FINAL.pdf (accessed 21 June 2016).

Latonero, M (2012), *The Rise of Mobile and the Diffusion of Technology-Facilitated Trafficking*, University of Southern California, available at https://technologyandtrafficking.usc.edu/files/2011/08/HumanTrafficking2012.pdf (accessed 23 June 2016).

Mendel, J and Sharapov, K (2014), 'Human trafficking and online networks', *Policy briefing*, Centre for Policy Studies, Central European University, Budapest, available at http://pasos.org/wp-content/uploads/2014/08/cps-policy-brief-upkat-human-trafficking-and-online-networks-2014.pdf (accessed 23 July 2016).

Mendel, J and Sharapov, K (2016), 'Human trafficking and online networks: policy, analysis, and ignorance', *Antipode*, vol. 48, no. 3, pp. 665–84.

Milivojevic, S (2012), 'The state, virtual borders and e-trafficking: between fact and fiction', in J McCulloch and S Pickering (eds), *Borders and Crime: Pre-Crime, Mobility and Serious Harm in an Age of Globalization*, Palgrave, Basingstoke, pp. 72–89.

Milivojevic, S and Pickering, S (2008), 'Football and sex: the 2006 FIFA World Cup and sex trafficking', *Temida*, vol. 11, no. 2, pp. 21–47.

Musto, J and boyd, d (2014), 'The trafficking-technology nexus', *Social Politics*, vol. 21, no. 3, pp. 461–83.

Nieuwenhuys, C and Pecoud, A (2007), 'Human trafficking, information campaigns, and strategies of migration control', *American Behavioral Scientist*, vol. 50, no. 12, pp. 1647–95.

O'Malley, P (2008), 'Neo-liberalism and risk in criminology', in T Anthony and C Cunneen (eds), *The Critical Criminology Companion*, Hawkins Press, Sydney, pp. 55–67.

Perer, A (2012), 'Policing the virtual red light district: a legislative solution to the problems of Internet prostitution and sex trafficking', *Brooklyn Law Review*, vol. 77, no. 2, pp. 823–59.

Pickering, S and Ham, J (2014), 'Hot pants at the border', *The British Journal of Criminology*, vol. 54, no. 1, pp. 2–19.

Piper, N and Roces, M (2003), *Wife or Worker? Asian Women and Migration*, Rowman & Littlefield, Oxford.

Romeo, D (2016), 'Hidden threat: the dark web surrounding cyber security', *Northern Kentucky Law Review*, vol. 43, no. 1, pp. 73–86.

Sarkar, S (2015), 'Use of technology in human trafficking networks and sexual exploitation: a cross-sectional multi-country study', *Transnational Social Review*, vol. 5, no. 1, pp. 55–68.

Segrave, M and Milivojevic, S (2015), 'Human trafficking: examining global responses', in G Barak (ed.), *The Routledge International Handbook of the Crimes of the Powerful*, Routledge, Abingdon Oxon, pp. 132–143.

Segrave, M, Milivojevic, S and Pickering, S (2009), *Sex Trafficking: International Context and Response*, Willan, Devon.

Siegel, R (2010), 'Craigslist shuts down international "adult services" sections', *Wired*, 18 December 2010, available at www.wired.com/2010/12/craigslist-adult-services-international/ (accessed 22 June 2016).

Sykiotou, A (2007), *Trafficking in Human Beings: Internet Recruitment*, Directorate General of Human Rights and Legal Affairs, COE, available at https://ec.europa.eu/anti-trafficking/sites/antitrafficking/files/trafficking_in_human_beings_internet_recruitment_1.pdf (accessed 24 June 2016).

Thakor, M and boyd, d (2013), 'Networked trafficking: reflections on technology and anti-trafficking movement', *Dialectical Anthropology*, vol. 37, no. 2, pp. 277–90.

UN.GIFT (2008), 'Background paper: 017 workshop – technology and human trafficking', The Vienna Forum to Fight Human Trafficking 13–15 February 2008, Austria Center Vienna, available at www.unodc.org/documents/human-trafficking/2008/BP017TechnologyandHumanTrafficking.pdf (accessed 23 June 2016).

USDOS (2011), *Trafficking in Persons Report*, available at www.state.gov/documents/organization/164452.pdf (accessed 21 June 2016).

Vance, C (2011), 'States of contradiction: twelve ways to do nothing about trafficking while pretending to', *Social Research*, vol. 78, no. 3, pp. 933–49.

Vanderschaaf, V (2013), 'How the internet facilitates underage victimization in human trafficking', *Children's Legal Rights Journal*, vol. 34, no. 1, pp. 135–8.

Vitale, K (2012), 'Barricading the information superhighway to stop the flow of traffic: why international regulation of the Internet is necessary to prevent sex trafficking', *American University International Law Review*, vol. 27, no. 1, pp. 91–131.

Wall, D (2013), 'Criminalising cyberspace: the rise of the internet as a "crime problem"', in Y Jewkes and M Yar (eds), *The Handbook of Internet Crime*, Willan, Devon, pp. 88–103.

Weber, L (2014), 'Peace at the border: a thought experiment', in L Weber (ed.), *Rethinking Border Control for a Globalising World: A Preferred Future*, Routledge, London, pp. 1–14.

Weitzer, R (2005), 'New directions in research on prostitution', *Crime, Law and Social Change*, vol. 43, pp. 211–35.

Weitzer, R (2012), 'Sex trafficking and the sex industry: the need for evidence-based theory and legislation', *Journal of Criminal Law and Criminology*, vol. 101, no. 4, pp. 1337–70.

Wired (2015), 'The escort database that combats human trafficking', 19 May 2015, available at www.wired.co.uk/article/human-trafficking-data-hack (accessed 23 June 2016).

Zhang, S (2009), 'Beyond the "Natasha" story: a review and critique of current research on sex trafficking', *Global Crime*, vol. 10, no. 3, pp. 178–95.

3 Feminist flight and fight responses to gendered cyberhate

Emma A. Jane

Introduction

> Layla[1] is a PhD student and member of a Facebook mother's page concerned that 'very nice' women such as the English academic Mary Beard are being attacked via the internet and social media platforms. As such, Layla is taking steps to keep her profile low in an attempt to dodge similar vitriol. Like many other women, this includes opting out of certain discussions, carefully moderating her tone, and ditching hashtags she knows cause grief. Much to her regret, Layla has also decided to 'unfollow' high-profile feminists she admires – feminists such as the UK activist Caroline Criado-Perez – because she can no longer stomach the toxic abuse filling their social media feeds (and consequently her own feeds) day after day. The 36-year-old mother of two is softly spoken, unfailingly polite, and – on the subject of her political persuasions – a self-described 'eternal moderate'. Asked what she thinks can be done to address the problem of misogyny online, Layla's tone remains as polite and as even as ever as she replies: 'Smash the patriarchy'.

This anecdote illustrates a seemingly paradoxical aspect of the increasingly serious international problem of cyber violence against women and girls (cyber VAWG): namely, that gendered cyberhate can be seen as provoking both 'flight' and 'fight' responses. As I will demonstrate over the course of this chapter, women are making changes to the places they go, the debates they join, the material they post and – in Layla's case – even the feminists they publicly support. Some are adopting male personas (to the extent of using voice modulating software while gaming) to avoid sexual harassment, while others are choosing to withdraw partly or wholly from the internet altogether (Jane 2017a, pp. 66–75). Yet alongside these metaphorical forms of 'flight' are increases in the visibility and force of women's push-back against misogyny online – what I am referring to as the 'fight' response. This includes an amplification of feminist rhetoric and increased individual and collectivist feminist activism, including online vigilantism or 'digilantism'.

'Digilantism' is a portmanteau used in both scholarly and media contexts to refer to acts that are illegal or legally liminal, as well as practices that are entirely legal and therefore more akin to traditional activism. Feminist digilante responses to gendered cyberhate most commonly involve 'calling out' or 'naming and shaming' attackers and can be situated on a spectrum. At the mildest end, activists republish material in a form that draws attention to the abuse but does not identify the perpetrators. Mid-range digilantism of this type utilises similar methods but also exposes at least the online identities of individuals. At the most extreme end of the naming and shaming spectrum, digilantes deploy tactics such as tracking down assailants offline and exposing them to the broader public, sometimes ensuring that perpetrators are 'outed' to family members, friends and employers, and inciting others to join the counter-attack (Jane 2016a, 2016b, 2017c).

This chapter draws on data from an ongoing series of research projects dedicated to mapping and studying the history, manifestations, nature, prevalence, aetiology and consequences of gendered cyberhate. While my formal research into this topic commenced in 2011, I have been archiving reports, self-reports and examples of gendered cyberhate since 1998 using methodological approaches from internet historiography (Brügger 2010). Since 2015, my research into the impact of cyberhate on the way in which women use the internet has received funding from the Australian federal government.[2] A key element of this part of my project has involved in-depth, semi-structured, qualitative interviews with 51 targets of gendered cyberhate.[3] I interviewed these women – aged between 19 and 52 – between 2015 and 2017. In this chapter, I offer analysis informed by those interviews, but I also provide two detailed case studies from the interview cohort, as they illustrate the range of women's experiences.

In previous work, I have used the term 'e-bile' to capture an array of discourses and practices that have historically been designated via terms such as 'cyberbullying', 'cyberstalking', 'trolling' and 'flaming' (Jane 2014a, 2014b, 2015). In this chapter, I use the general terms 'gendered cyberhate', 'gendered e-bile' and 'cyber VAWG' interchangeably to refer to a range of discourses and acts occurring at the gender–technology–violence nexus. These include sexually violent invective, plausible rape and death threats, stalking, large groups attacking individuals, the malicious circulation of targets' personal details online ('doxing'), and the uploading of sexually explicit material without the consent of the pictured subject ('revenge porn'). My approach to terminology is intended to recognise the complexity and interrelated nature of the acts under analysis and the fluidity and changing nature of these practices and the technologies on which they rely. In this chapter, I use the term 'target' rather than 'victim' or 'survivor' in relation to the language used to describe those who have experienced sexual violence. This is in recognition of feminist debates about the way in which 'victim' connotes helplessness and 'survivor' can connote a type of 'forced heroism' (Sehgal 2016). I do, however, use the expression 'victim-blaming' for idiomatic reasons (that is, because the term 'victim-blaming' has connotations that 'target-blaming' does not). The reader should note that my focus on gendered cyberhate should not be read as denying or downplaying online hate speech that is racist, homophobic, transphobic and so on.[4]

In this chapter, I detail the nature, prevalence and ramifications of cyber VAWG, as well as the manifest failures of police, policy makers and platform managers to intervene in a manner that adequately supports targets, punishes perpetrators and/or assists in addressing the broader problem of cyber VAWG at all jurisdictional levels. This situation is shown to be a source of great frustration and anger for women and helps explain why many feel that they have no option but to take matters into their own hands. I then outline some of the ways in which individual women are navigating the problem in their own daily practices and their bigger-picture activism. Two ostensibly contrasting case studies involving high-profile Australians – one whose response might be characterised as involving 'flight' and one who has engaged in a more obvious 'fight' – are offered by way of illustration. As I will show, however, elements of both flight and fight are observable not only within feminist responses to gendered cyberhate collectively but at the individual level as well.

'You can't get called a cunt day in, day out for 10 years and not have that make a really serious impact on your psyche': the harm of gendered cyberhate

Gendered cyberhate was relatively rare and mild in the early decades of the internet but has become far more prevalent, visible, noxious and directly threatening since at least 2010 (Jane 2017a, pp. 16–42). These amplifications are likely a flow-on effect from the self-publishing and networking opportunities associated with what is known as the Web 2.0 era. ('Web 1.0' is generally used to describe those early decades of the internet when content was mostly static and delivered in a read-only format. 'Web 2.0' refers to the shift – most obvious from around 2006 – towards user-generated material, interactivity, collaboration and sharing.) Put simply, the Web 2.0 era has given online antagonists access to targets (and appreciative audiences) in a way that was not previously possible. Gendered cyberhate typically involves discourse that passes scathing and explicit judgement on women's appearance, sexual attractiveness and/or perceived sexual activeness; deploys ad hominem invective; is couched in terms involving hyperbolic misogyny, homophobia and/or sexually graphic imagery; prescribes coerced sex acts as all-purpose correctives; and/or demonstrates disavowal in that authors fixate on their targets while insisting explicitly or implicitly that nothing about their targets is worthy of notice.

Numerous anecdotal reports of gendered cyberhate are supplemented by emerging statistical data. A 2015 United Nations (UN) report, for instance, synthesises a range of empirical research to show that 73 per cent of women and girls have been exposed to or experienced some form of 'online violence' (UN Broadband Commission for Digital Development Working Group on Broadband and Gender [UN Broadband Commission] 2015, p. 2). Further, the report states that women are 27 times more likely to be abused online than men, that 61 per cent of online harassers are male, and that women aged between 18 and 24 are at particular risk (UN Broadband Commission 2015, p. 15). In 2016, meanwhile, *The Guardian* conducted a revealing analysis of 70 million remarks posted on its own comment threads. Researchers isolated those comments blocked for violating the organisation's community standards (usually because they were regarded as 'abusive or disruptive' [Mansfield 2016]) to determine that, of the 10 regular writers who received the most abuse, eight were women (Gardiner et al. 2016). The writer who attracted the largest number of objectionable comments was the feminist commentator Jessica Valenti (Valenti 2016).

Gendered cyberhate has the potential to cause emotional, social, financial, professional and political harm, in that – among other consequences – it can constrain women's ability to find jobs, market themselves, network, socialise, engage politically and partake freely in the sorts of self-expression and self-representation regarded as key benefits of the Web 2.0 era. Career derailment and financial loss are not only incidental by-products of gendered cyberhate but may also come about as a result of deliberate individual or group attempts to sabotage the jobs and ongoing employment prospects of targets (Jane 2017a, pp. 66–8).

While cyberhate need not involve offline dimensions to cause "embodied harm" (Henry and Powell 2015), the increasing number of attacks that combine online and offline dimensions highlights the fact that cyberhate targets may well be at risk of suffering bodily harm (as it is understood in the traditional sense). Doxing, for instance, is often accompanied by incitements to internet antagonists to hunt targets offline. A frequently cited case study involves a man who posted an ad titled 'Rape Me and My Daughters' which included his ex-wife's home address and prompted more than 50 strangers to arrive at this woman's home (Sandoval 2013).

Doxing is particularly harmful to women when combined with revenge porn. In 2011, Holly Jacobs, a graduate student and tutor at a Florida university, received a tip that explicit images she had shared with a former boyfriend had been posted online onto hundreds of revenge porn sites alongside an explanation of how to locate her offline and an expression of interest – made in her name – in sex with strangers (Citron 2014, pp. 45, 48). As a result of the attacks, Jacobs bought a stun gun, changed her name, withdrew from online activities and stopped walking alone at night because she felt she had become 'a prime target for actual rape' (cited in Citron 2014, pp. 46, 48). Yet Jacobs struggled to obtain adequate support from law enforcement and platform operators. She was turned away by two different police departments as well as the Federal Bureau of Investigation. Of the multiple websites she approached to have the images removed, several said that they would do so but only for a fee (Citron 2014, pp. 46–7). Jacobs's experience highlights the way in which online sexual violence (like the offline version) inflicts many layers of suffering, in that women often suffer not only the initial assault but also victim-blaming and sex-related shaming in the aftermath. She was also placed in the position of having to bear both the metaphorical and literal costs of seeking remedy for the revenge porn attacks against her.

Other particularly violent forms of gendered cyberhate involve emerging practices such as rape video blackmail and 'sextortion'. In Australia (as in other nations), anecdotal accounts suggest that a growing number of women are being raped and then blackmailed with video footage of the attack (Selvaratnam 2015). 'Sextortion', meanwhile, involves obtaining intimate footage – usually by hacking computers and webcams remotely – and then using this footage to blackmail targets into performing sexual acts online (Wittes et al. 2016). One sextortionist arrested by federal investigators in the United States (US) was found with more than 15,000 webcam video captures, including videos he surreptitiously recorded showing targets getting out of the shower and having sex with their partners. This perpetrator had 900 audio recordings and 13,000 screen captures, possessing files associated with 129 computers and roughly 230 people, including 44 minors (Wittes et al. 2016, p. 2).

The impact of gendered cyberhate varies depending on individual contexts and the severity of the abuse. In its study into sextortion, the Brookings Institution found multiple cases of targets contemplating, threatening or even attempting suicide (Wittes et al. 2016, p. 5). One Australian woman subject to rape video blackmail describes being so upset that she vomited when she first realised that her sexual assault had been filmed (cited in Selvaratnam 2015).

Commenting on the impact of the ongoing abuse she is subjected to, Valenti says, 'You can't get called a cunt day in, day out for 10 years and not have that make a really serious impact on your psyche' (cited in Goldberg 2015). As such, she has stopped promoting her speaking events publicly and started hiring security (Hess 2014). She says she feels like quitting her feminist activism 'all the time' (cited in Goldberg 2015). In 2016, Valenti announced that she was withdrawing from social media because a rape and death threat had been directed at her five-year-old daughter (Morris 2016).

These comments comport with my own findings. Many of the women I have interviewed demonstrate great resilience in terms of being able to matter-of-factly rationalise and sometimes even poke fun at the abuse and harassment they experience online. However, my observations also reveal that (1) many targets reach one or more emotional breaking points when gendered cyberhate is particularly severe, hits a particularly raw nerve or reaches a critical mass; and (2) the cumulative, long-term effect of dealing with such material day in and day out can be insidious for women's psychological state, as well as their mode of engagement in both the online and offline public spheres. My research findings[5] show that gendered cyberhate targets who reach breaking point often describe feeling some combination of intense anxiety and fear and/or report mental health issues such as depression, anxiety disorder, agoraphobia and panic attacks. Insomnia is common, as are social withdrawal and feelings of paranoia about whether people encountered offline are online attackers. Some women seek professional medical assistance, have been diagnosed with significant mental health disorders and are prescribed medication.

In addition to harming individuals, gendered cyberhate poses a threat to broader ideals and political goals such as those relating to online equity and digital citizenship. Yet despite gendered cyberhate potentially resulting in significant internet exclusion and under-use (Jane 2017b), police, policy makers and platform operators have mostly failed to respond in a useful and/or appropriate manner. That is, they have not responded in a manner that recognises the seriousness of cyber VAWG; supports women in obtaining relief from and redress for attacks; brings perpetrators to account; and scrutinises (and where necessary intervenes in) platform design, governance and complaint-handling protocols.

Anecdotal complaints by women are corroborated by statistical data. For example, a World Wide Web Foundation report shows that, in 74 per cent of Web Index[6] countries, law enforcement agencies and the courts are failing to take appropriate action in response to acts of gender-based violence online (nd, pp. 15, 4). A report by the Association for Progressive Communications (APC) identifies multiple policy failures at an international level in terms of addressing the increasing levels of violence against women enacted via information and communications technology (ICT) (2014). It finds that violence involving ICT is not being prioritised in prevention and response strategies, budgeting or evidence-based policy making, leaving female targets with little or no avenue for redress (APC 2014, p. 4). Comparing the policies of Facebook, YouTube and Twitter, the APC also identifies a reluctance to engage directly with technology-related violence against women until it becomes a public relations issue; a lack of transparency around reporting and redress processes; a failure to engage with the perspectives of non–North American/European women; and no public commitment to human rights standards or to the promotion of rights other than the encouragement of free speech (Nyst 2014, pp. 3–4). The UN, meanwhile, warns that if cyber VAWG remains unchecked, it risks producing 'a 21st century global pandemic with significant negative consequences for all societies in general and irreparable damage for girls and women in particular' (UN Broadband Commission 2015, pp. 6–7).

It is beyond the scope of this chapter to propose solutions for cyber VAWG or to discuss critical issues such as the intersection of criminal justice and service provider responsibility for first-line responses to and long-term prevention of gendered cyberhate. These are important lines of inquiry for future research. Suffice to note here only that the manifest failure of various authorities and institutions to address cyber VAWG has placed many women in a situation in which they feel that DIY responses are the only options available to them.

Towards a taxonomy of responses to gendered cyberhate

In this section, I detail some of the tactics women are using to navigate abuse and harassment in online domains. My research findings reveal that the practices women deploy in response to cyberhate can be grouped thematically into the following overlapping categories:

- Distancing: In forms such as ignoring, blocking, muting and deleting objectionable content and users.
- Rationalising: For example, explaining away abuse in ways that render it less personal, such as focusing on the fact that it is a systemic, gender-related issue.
- Technological 'hygiene'[7]: That is, setting personal boundaries around technology in an effort to reduce the impact of cyberhate while still accessing and fielding such material. (Examples include refraining from checking email and social media accounts after hours or in bed and outsourcing email and social media account-checking to friends or colleagues when attacks are underway or anticipated.)
- Restricting internet use: Includes strategies such as avoiding (where possible) interactions with strangers; retreating from contentious debates; steering clear of hashtags regarded as or previously experienced as contentious; removing or greatly restricting the circulation of personal information and photos; using the web only or predominantly in broadcast or 1.0 mode;

seeking refuge in private spaces such as closed Facebook groups; 'locking down' or privatising accounts; disabling comment sections on blogs; and/or taking short, long or permanent breaks from parts or all of the internet.

- One-on-one engagement with attackers: Examples include confronting and arguing with online antagonists either privately or publicly. This may involve responding with similarly abusive tactics and/or with reason, humour or good-natured appeals to attackers as fellow internet users.
- 'Traditional' activism: Includes awareness-raising and advocacy work such as writing/speaking about cyberhate, signing or launching petitions, lobbying platforms and policy makers, rallying support from online and/or offline communities, providing assistance to other female targets, forming or joining activist groups and archiving received cyberhate for later use.
- Performance-based activism: Strategies include the humorous and creative repurposing of cyberhate. An example is the increasingly common practice of 'performing' online hate speech by reading such material aloud in publicly circulated videos.
- Digilante activism: Engaging in extrajudicial practices that are intended to punish attackers or otherwise bring them to account.

Before looking more closely at digilantism, I offer the following two putatively contrasting case studies. These two women were among the 'public profile' participants approached for the research project described earlier and in footnote 3. Tracey Spicer was interviewed via Skype, while Jenna Price was interviewed in person. It is worth noting that Spicer has also reflected on her experience in her own public writing, and this is indicated in the case study via a reference; otherwise all quotes are from the interview. I have chosen to focus on these two case studies for two key reasons. First, while at first glance they may appear to involve starkly contrasting responses, closer inspection reveals that they actually have some commonalities. This leads to the second rationale, which is that these case studies demonstrate many of the practices detailed earlier. Spicer's responses, for instance, involve elements of distancing, rationalising, restricting internet use, traditional activism, performance-based activism and (eventually) digilante activism. Price, meanwhile, describes engaging in forms of distancing, rationalising, technological hygiene, one-on-one engagement with attackers, traditional activism and digilante activism.

Case study 1: Tracey Spicer

Tracey Spicer is an Australian newsreader, documentary maker, journalist and writer with a nearly three-decade career in the media. She began experiencing a large volume of online 'sexualised violence' after she became more political and overtly feminist in her media commentary around 2013. Since then, the vitriol directed at Spicer has arrived 'pretty much every day' and has included mob abuse by men's rights groups from around the world and threats to rape and murder Spicer and her children. A particularly savage attack unfolded in April 2014 after she wrote a piece in her weekly travel column saying that, if her children were travelling alone on a plane, she would prefer that they be seated next to a female rather than a male passenger because of the high percentage of male perpetrators in child abuse cases (Spicer 2014). This column, she says, 'really opened the floodgates'. She was on holidays in Vietnam with her family when she received:

> really vile attacks from all over the world . . . 'you deserve to be raped but you're too ugly. I wouldn't want to fuck your children anyway', this kind of just horrific stuff . . . I got quite scared . . . I was actually frightened to return to my own home . . . because the threats were so horrific – not only to me, but to my children. It gave me an awful fright . . . some of the ones that were from the men's rights groups in America were quite explicit . . . 'we know that you're in Australia, but we know where your kids go to school', this kind of stuff . . . there were absolutely direct threats to kill me and to rape me and to kill the children.

When Spicer rang her employers from Vietnam, she discovered that her editors were excited that her column was 'really going off' in that it was attracting a large number of comments. When she explained that she was 'absolutely terrified' by the response, her employers complied with her request to move the story off the front page of their website. Spicer remembers feeling suffocated:

> While I'm not the kind of person who's prone to anxiety . . . I did go through months where I was more housebound. I didn't want to go outside that much. I looked over my shoulder while I was outside. . . . When they pile upon you, you feel like you don't want to go out the front door because there might be a mob out there. That's how you feel. It almost manifests physically in your mind that there are people out there with pitchforks.

Spicer decided against reporting the attacks to police because she was worried that this would result in more and worse attacks on both herself and her children. She said that she lacked the confidence to follow the lead of other feminists who had republished the abuse they received – although she said that she had great admiration for the courage of these women. Her preferred strategy is still to 'mute or block' anyone who attempts to engage with her by using violent language. She has also changed both the content and style of her writing: 'I'm a little bit ashamed to admit this, but I've been frightened to write too many full-on columns ever since then . . . I've really eased off on writing edgy columns because it scared the shit out of me'.

In many ways, this case study highlights the 'flight' reaction, which seems to contrast with Price's 'fight' response outlined next. As I will go on to explain, however, Spicer and Price – like many other gendered cyberhate targets I have interviewed – have toolkits of responses that are deployed in a context-specific manner.

Case study 2: Jenna Price

Jenna Price is an Australian academic, former journalist and co-founder of the controversial Destroy the Joint project, whose anti-violence campaign is called 'Counting Dead Women'. Price frequently receives large volumes of gendered cyberhate – including threats of sexual violence towards her daughters – to which she often responds with digilante tactics. From late 2012 to early 2013, for instance, she was attacked by a man – 'David'[8] – on Twitter who, in addition to calling her a 'slut', a 'whore' and a 'cunt', accused her of being a liar and thief who had stolen money from a union and used it to send her children to private schools. Price tracked down this man's identity and rang the café at which he worked. When his wife answered the phone, Price was 'very brutal' in telling her about the messages that she had received from her husband. After asking to speak to her husband, she said:

> 'Hi, it's Jenna Price'. There was silence. I said, 'You're calling me all these things. You're wrong. I'm going to call a lawyer if you do it again . . . do you understand?', and he went, 'yeah', and . . . I hung up.

Price then decided to follow up with a visit to the café in person. When asked why, she explained that it was because she's 'just that kind of person':

> I went to the café and I said, 'My name's Jenna Price, we've had a phone conversation, do you understand what I said to you?' He went, 'yeah' . . . I have never heard a word from him since that time,

> and that's been fucking awesome . . . I was really angry . . . I was so angry . . . I was just really wanting to frighten him in the same way that I had felt frightened.

Price has been similarly proactive on other occasions, relying on the skills she learned in her previous career as a journalist to track down online abusers at their workplaces. On one occasion, she used an IP address to deduce that 'absolutely appalling emails' were coming from someone employed within a mid-level Australian telecommunications company. Unable to identify the individual involved, she rang the organisation's chief executive officer (CEO), explained the sorts of material she was receiving and said that she would seek legal action unless they ceased. When asked about the response that she received from this CEO, Price described it as 'Fantastic: terror, complete terror'.

Flight and fight

The Spicer and Price case studies highlight two thematic modes of gendered cyberhate responses. While 'flight' is too strong a word to describe Spicer's actions, she did withdraw in terms of staying in her house more often than usual, as well as self-censoring her writing and muting the public expression of her political views. Her suffering was multilayered in that, in addition to having her freedom of expression restricted and being extremely frightened, she also punished herself for deciding to tone down her writing. It is interesting to note, however, that – from an outside perspective – Spicer seems to have remained politically active and has continued to speak in an upfront and potentially controversial way about many social issues. With regards to gendered cyberhate, for example, she has engaged in activism that includes reading abuse she has received online to a camera ('Women Read Mean Tweets' 2015), as well as publicly supporting an en masse, feminist 'naming and shaming' operation which played out on Facebook and Twitter in early 2016 (Tay 2016).

Spicer's ongoing activism demonstrates three things. First, it shows that outsiders may not be able to accurately appraise whether or by how much a woman's voice and politics are being muted as a result of gendered cyberhate. While Spicer still seems politically to be extremely active, by her own assessment, she is not as 'edgy' as she used to be, at least in her writing. Second, it shows the way in which individual women's responses to gendered cyberhate cannot always be neatly categorised as one thing or another. Spicer feels that she has pulled back in terms of her writing style and content ('flight') yet is still participating in overt feminist activism including digilantism ('fight'). Third, it illustrates a point I will make in greater detail presently – namely, that celebratory media coverage of feminist digilantism in response to gendered cyberhate may imply that women who respond in other ways lack courage or conviction or are somehow letting the feminist team down. This, in turn, could contribute to the sense of embarrassment and self-shaming described by Spicer.

Price's response to gendered cyberhate clearly involves a 'fight'. That said, like Spicer, her responses are complex and change over time. For instance, when she first became involved with Destroy the Joint and received a great deal of abuse in 2010, Price says that she used to get 'very upset' – crying, suffering from sleeplessness and anxiety, and leaning on her partner for emotional support. Her current mindset, she says, has come about because, after a year or so of abuse, she hardened and developed a 'fuck you all' attitude. In relation to the incident in which she confronted her attacker at his café, she also recalls acting out of a place of deep frustration: 'I did it because I had lost my mind'. Further, digilantism is just one of the approaches Price uses to navigate gendered e-bile. She also deploys a multitude of filters; is a fan of block, mute and delete options on social media platforms; no longer checks her accounts on her phone in her bedroom; and sometimes asks male friends to respond to antagonists on her behalf. It is revealing that, while Price demonstrates pride and satisfaction in the results of her own digilantism, she does not recommend that other women, especially younger, more vulnerable women, follow suit.

Feminist digilantism

While Price's tactics are an unusually strong form of digilantism, similar strategies are increasingly being deployed by other feminists and, for the most part, are receiving extremely celebratory media coverage (Jane 2016b). In late 2014, for instance, the Australian gamer Alanah Pearce began contacting the mothers of boys who were sending her rape threats on Facebook and then publishing these digilante efforts on Twitter (although not in a way that identified the perpetrators or their families). Pearce's tactics received rhapsodic international media coverage and were repeatedly described as the 'perfect' way for women to deal with rape threats online (cited in Jane 2016b). Given the over-simplistic and arguably sensationalist tone of such reporting, a more sober, scholarly appraisal of digilantism is useful.

In the introduction of this chapter, I explained that naming and shaming digilantism exists on a spectrum. An example from the more extreme end occurred in 2012 when the Canadian feminist Steph Guthrie unearthed the identity of the 25-year-old man who created an online game inviting players to punch an image of the feminist commentator Anita Sarkeesian so that Sarkeesian's face became increasingly bruised and damaged. Guthrie confronted the man on Twitter, asking him whether he also punched women in the face 'IRL' (that is, 'in real life'). Determined that this man face offline consequences, she also passed his details on to media outlets, as well as alerting potential employers in his local area of his activities, so that if he ever sent them his résumé, they would know that he was responsible for making 'woman facepunching' games (Guthrie 2012, 2013).

In Australia in December 2015, meanwhile, the writer and feminist Clementine Ford tagged a man's employer into a re-post of an abusive comment that he had made about her on Facebook. This man was subsequently sacked (Levy 2015). Also in Australia that year, a group of young women who subsequently formed the advocacy group Sexual Violence Won't Be Silenced (SVWBS) engaged in multifaceted activism in response to the 'slut shaming'[9] of one of their friends on Facebook. While their activism began with digilantism that led to one man losing his job, it also involved reporting the online rape threats made to various women to police (Jane 2017c). As a result, a 25-year-old man pleaded guilty to using a carriage service to menace, harass or cause offence and, in July 2016, was placed on a 12-month good-behaviour bond (Jane 2017c).

Feminist digilantism offers a number of benefits. At the individual level, it can return an empowering sense of agency. Collectively, it can raise public awareness about individual incidents as well as the broader problem of misogyny online. Such tactics can also hold perpetrators accountable for their actions when institutions have failed to do so. In the SVWBS case, we can also see that efforts that begin as digilantism may trigger regulatory responses, as well as stimulating more traditional, collectivist activism. That said, as with privately sourcing criminal justice via 'violent self-help' in offline domains (Hine 1998, p. 1223), online vigilantism has many downsides.

Here it is useful to look to the legal literature dealing with offline vigilantism. A number of scholars writing in this area make persuasive cases that, while vigilantism, by definition, can never be legally justified, in some circumstances it may not only be morally *justified* but morally *demanded* (Robinson 2015, p. 405). The argument is that a 'justified vigilantism' defence should be available in the criminal justice system, where deficits in public law enforcement mean that society is failing to hold up its end of the social contract (Hine 1998, pp. 1252–3). Convincing cases have also been made in relation to female vigilantism in contexts such as domestic violence (Ayyildiz 1995, pp. 146–7), as well as nations like India where, it is argued, vigilantism is a 'social necessity' because of systemic and oppressive gender inequity (Sen 2012; White and Rastogi 2014).

That said, the outcomes of digilantism are unpredictable and may backfire in a manner that disadvantages activists. Extrajudicial activism online also provides opportunities for bad actors who may join digilante salvos for recreational purposes rather than for reasons relating to politics or self-defence. This is especially true in pack attacks in which some or even many participants have little association with or interest in the original offender or offence yet still coalesce into large, hostile cyber mobs (see Citron 2014, p. 118; Ronson 2015). Crowds of angry feminist digilantes who attack and dox individual men risk mirroring the very behaviour being objected to in the first instance – even if their motivations are different and more defendable. Digilantism may also undermine confidence in justice systems in a manner that promotes further bypassing of legal channels rather than advocacy for the sorts of reform that might prevent or reduce such attacks on women. In other words, while digilantism may ultimately strengthen extrajudicial cultures online, my case is that more rather than fewer regulatory responses are required.

Given the popularity of fictional vigilante crime fighters in mass culture (a recent example is Jessica Jones from the television series *Jessica Jones* [Iacofano

2015]), it is not surprising that – despite the many downsides listed – real-life feminist digilantes receive such an exuberant mainstream media reception. There is a risk, however, that gushing media coverage reifies the view that no other interventions are required with regards to the gendered cyberhate situation. Further, media applause for the actions of feminist digilantes as constituting the 'perfect' solution to online rape threats comports with the wider trend of shifting the burden of responsibility for the problem of gendered cyberhate from male perpetrators to female targets and from the public to the private sphere (Jane 2016b). Noteworthy, for instance, is the way in which the Pearce case involved an all-female clean-up crew, in the form of the female target and the perpetrators' mothers. In such scenes, mothers, wives and girlfriends are propelled into the role of domestic 'police' who must hold their menfolk to account. Indeed, in this example, it is a woman who ends up doing the apologising.[10]

In addition to placing a monumentally unfair burden on targets, approaches such as Pearce's are unfeasible in the vast bulk of cyberhate cases. Attacks usually involve a multitude of assailants, which would make it prohibitively time-consuming for women to track down and confront them all. This task would also be difficult if not outright impossible in the case of antagonists attacking anonymously or pseudonymously (although, as I explain in what follows, it is becoming harder to use the mainstream domains of the cybersphere anonymously). We should also remember that individual women might not fare so well in one-on-one confrontations with perpetrators if they happen to be subordinate to these men in their social circumstances and relationships.[11] Finally, all-out celebrations of digilantism may also create the expectation that this is the best or 'correct' feminist response to gendered cyberhate which, in turn, may contribute to the suffering of women – like Layla and Spicer – who, in some circumstances, choose to withdraw in various ways rather than engage in overt confrontation and battle.

Before concluding this section, it is worth noting the complex and dynamic relationship between feminist digilante responses to gendered cyberhate and anonymity online. Anonymity is often assumed to be one of the prime reasons why the internet has become so saturated with vitriol (Hess 2014). Yet the culture of anonymity online is changing. In this regard, the reader might note Facebook's growing push for users to sign up with their 'authentic identities' (Facebook nd) and the fact that increasing numbers of third-party applications (including the dating app Tinder) are making it mandatory for users to register via Facebook, thereby importing this identifying profile information. These sorts of 'real name' policies are making it increasingly difficult to use mainstream sectors of the cybersphere in an anonymous manner. Thus, while some feminist digilantes have had to use a degree of sleuthing to hunt down their cyber antagonists, others have simply directed public attention to their attackers' identifying Facebook profiles. The growing ease with which feminists are able to identify their online attackers can therefore be seen as a double-edged sword. On the one hand, it is becoming easier to publicly expose perpetrators. On the other hand, men do not seem overly concerned about being exposed in this way. This lends support to the case that gendered cyberhate has become normalised to the extent that many assailants do not anticipate criminal sanction, platform intervention or even social stigma for their actions.

Conclusion

In this chapter, I have outlined the way in which women's responses to the growing problem of gendered cyberhate can involve both flight and fight modes. The former has a chilling effect in that online abuse and harassment can cause women to restrict their use of the internet and engagement online in ways that are disempowering and disenfranchising. The latter manifests in the individual and collective push-back against gendered cyberhate, including some relatively forceful digilante strategies. Throughout this chapter, I have endeavoured to avoid simplistic framings of gendered cyberhate targets as either silenced 'victims' or romanticised guerrilla warriors. Instead, I have shown that women who are attacked and harassed online respond in a range of ways, using different strategies to suit different contexts. Further, strategies associated with both flight and fight modes have upsides and downsides.

A central aim of this chapter has been to discourage appraisals of various forms of feminist activism in terms of their potential to 'fix' the broader problem of cyber VAWG. Women's range of responses to gendered cyberhate is eminently understandable given the surplus of abuse and the deficit of institutional assistance, and there is much to praise about targets' courage, savvy and innovation. Yet acknowledging women's skill sets in this regard should not obscure the fact that these are skill sets that women have had to develop because of profoundly unfair and oppressive circumstances. Misogyny online (like racism and homophobia online) is the result of far broader patterns of subordination and violence. In this regard, the various approaches discussed in this chapter illustrate the resilience not only of feminist resistance to various forms of gendered violence but also of the structural inequity that facilitates exactly the sorts of gendered violence feminists are resisting. As such, any discussion of these matters should keep sight of the fact that all the praiseworthy feminist resistance would be unnecessary if it were not for all the lamentable patriarchal oppression.

In particular, the fact that women are finding innovative ways to navigate gendered cyberhate should never be used, either explicitly or implicitly, to relieve regulators and policy makers of their responsibilities with regards to cyber VAWG. As I have argued, the risks and downsides of the range of strategies described in this chapter should add further pressure to policy makers and police to intervene in a manner that moves the focus away from the female targets of gendered cyberhate and towards the multitude of male perpetrators who are currently abusing and harassing with impunity, often on platforms whose operators show little interest in protecting vulnerable users. This might bring us closer to the day when we are no longer celebrating women's impressive abilities to navigate the deluge of gendered cyberhate online but are celebrating the fact that these abilities are no longer required.

Notes

1 'Layla' is a pseudonym.

2 This is in the form of a Discovery Early Career Researcher Award (DE150100670) which is funding a three-year project called 'Cyberhate: the new digital divide?' The University of New South Wales (UNSW) Research Ethics Committee reference for this project is HC15012.

3 Two groups of interviewees were recruited via a number of methods. The first group (n = 32) comprised women with public profiles who had experienced hostility or threats online and had previously discussed this in a public forum. These women had the option of being interviewed in an identifiable way using their real names, and most made use of this option. The second group (n = 20) comprised women who were not in public life and who had experienced hostility or threats online but had not spoken about this previously in a public forum. These interviewees all used pseudonyms, and their identifying details were removed from their transcripts. While my recruitment techniques were not designed to obtain a representative population sample, I did ensure that my subjects included women of colour, queer women and Muslim women, as well as women from a range of age groups and socioeconomic circumstances. Given that this chapter cites women from my interview cohort as well as women speaking in other contexts, I will indicate throughout where subjects were interviewed by me personally and also when pseudonyms are used. Interviews were conducted in person or via Skype, with some follow-up interviews involving phone conversations and email.

4 While the political intersectionality of gender with other social identities is indubitable, it is simply beyond the scope of my current research to investigate in any detail cyberhate as it relates to issues such as race, class, sexual orientation and gender. Related to this is the fact that, while some international statistics are cited, the qualitative dimensions of my work are entirely Anglophone.

5 These findings are drawn from my interviews with cyberhate targets, as well as from my analysis of my archive of reports and self-reports of gendered cyberhate.

6 The World Wide Web Foundation's Web Index covers 86 countries – including many high-income nations – and measures the web's contribution to social, economic and political progress.

7 I am using the term 'hygiene' here in line with the use – by various clinicians – of the term 'sleep hygiene' to refer to practices and household arrangements (for example, removing clocks from bedrooms) designed to facilitate quality sleep.

8 'David' is a pseudonym.

9 'Slut shaming' is a colloquial term describing 'the act of criticizing women or girls for their real or presumed sexuality or sexual activity, as well as for looking or behaving in ways that are believed to transgress sexual norms' (Karaian 2014, p. 296).

10 Many thanks to Melanie Andersen for her input into this section.

11 Many thanks to Melanie Andersen for her input into this section.

References

Association for Progressive Communications (2014), *Domestic Legal Remedies for Technology-Related Violence Against Women: Review of Related Studies and Literature*, May, available at www.genderit.org/sites/default/upload/domestic_legal_remedies_for_technology-related_violence_against_women_review_of_related_studies_and_literature.pdf (accessed 28 January 2016).

Ayyildiz, E (1995), 'When battered woman's syndrome does not go far enough: the battered woman as vigilante', *Journal of Gender & the Law*, vol. 4, no. 141, pp. 141–66.

Brügger, N (ed.) (2010), *Web History*, Peter Lang, New York.

Citron, DK (2014), *Hate Crimes in Cyberspace*, Harvard University Press, Cambridge, Massachusetts, and London, England.

Facebook (nd), *What Names Are Allowed on Facebook?* available at www.facebook.com/help/112146705538576 (accessed 21 September 2016).

Gardiner, B, Mansfield, M, Anderson, I, Holder, J, Louter, D and Ulmanu, M (2016), 'The dark side of Guardian comments', *The Guardian*, 12 April, available at www.theguardian.com/technology/2016/apr/12/the-dark-side-of-guardian-comments (accessed 13 April 2016).

Goldberg, M (2015), 'Feminist writers are so besieged by online abuse that some have begun to retire', *The Washington Post*, 20 February, available at www.washingtonpost.com/opinions/online-feminists-increasingly-ask-are-the-psychic-costs-too-much-to-bear/2015/02/19/3dc4ca6c-b7dd-11e4-a200-c008a01a6692_story.html (accessed 14 January 2016).

Guthrie, S (2012), 'Why is this conversation necessary?', *Storify*, available at https://storify.com/WiTOpoli/why-is-this-conversation-necessary-ben-spurr (accessed 19 March 2016).

Guthrie, S (2013), 'The problem with "Don't Feed the Trolls"', *TEDxToronto*, 1 October, available at www.youtube.com/watch?v=_KHEkR5yb9A (accessed 19 March 2016).

Henry, N and Powell, A (2015), 'Embodied harms: gender, shame, and technology-facilitated sexual violence', *Violence Against Women*, vol. 21, no. 6, pp. 758–99.

Hess, A (2014), 'Why women aren't welcome on the internet', *Pacific Standard*, 6 January, available at www.psmag.com/health-and-behavior/women-arent-welcome-internet-72170 (accessed 17 December 2015).

Hine, KD (1998), 'Vigilantism revisited: an economic analysis of the law of extra-judicial self-help or why can't Dick shoot Henry for stealing Jane's truck?' *The American University Law Review*, vol. 47, pp. 1221–54.

Iacofano, T (Producer) (2015), *Jessica Jones*, Netflix, New York City.

Jane, EA (2014a), '"Your a ugly, whorish, slut": understanding e-bile', *Feminist Media Studies*, vol. 14, no. 4, pp. 531–46.

Jane, EA (2014b), '"Back to the kitchen, cunt": speaking the unspeakable about online misogyny', *Continuum: Journal of Media & Cultural Studies*, vol. 28, no. 4, pp. 558–70.

Jane, EA (2015), 'Flaming? What flaming? The pitfalls and potentials of researching online hostility', *Ethics and Information Technology*, vol. 17, no. 1, pp. 65–87.

Jane, EA (2016a), '"Dude . . . stop the spread": antagonism, agonism, and #manspreading on social media', *International Journal of Cultural Studies*, published online 10 March, DOI: 10.1177/1367877916637151.

Jane, EA (2016b), 'Online misogyny and feminist digilantism', *Continuum: Journal of Media & Cultural Studies*, published online 31 March, DOI: 10.1080/10304312.2016.1166560.

Jane, EA (2017c – forthcoming, accepted for publication on 2 September 2016), 'Feminist digilante responses to a slut-shaming on Facebook', *Social Media + Society*.

Jane, EA (2017a), *Misogyny Online: A Short (and Brutish) History*, Sage, London.

Jane, EA (2017b – forthcoming, accepted for publication on 30 July 2016), 'Gendered cyberhate: a new digital divide?', in M Ragnedda and GW Muschert (eds), *Theorizing Digital Divides*, Routledge, Oxon.

Karaian, L (2014), 'Policing "sexting": responsibilization, respectability and sexual subjectivity in child protection/crime prevention responses to teenagers' digital sexual expression', *Theoretical Criminology*, vol. 18, no. 3, pp. 282–99.

Levy, M (2015), 'Hotel worker Michael Nolan sacked over Facebook post to Clementine Ford', *The Sydney Morning Herald*, 1 December, available at www.smh.com.au/national/hotel-worker-michael-nolan-sacked-over-facebook-post-to-clementine-ford-20151130-glcly4.html (accessed 15 July 2016).

Mamamia (2015), 'Women Read Mean Tweets', 23 February, available at www.youtube.com/watch?v=Xs4GC2WCy3U (accessed 15 July 2016).

Mansfield, M (2016), 'How we analysed 70m comments on the Guardian website', *The Guardian*, 12 April, available at www.theguardian.com/technology/2016/apr/12/how-we-analysed-70m-comments-guardian-website (accessed 21 September 2016).

Morris, DZ (2016), 'Bestselling feminist author Jessica Valenti quits social media after rape and death threats directed at daughter', *Fortune*, 31 July, available at http://fortune.

com/2016/07/31/bestselling-feminist-author-jessica-valenti-quits-social-media-after-rape-and-death-threats-directed-at-daughter/ (accessed 25 September 2016).

Nyst, C (2014), *End Violence: Women's Rights and Safety Online Internet Intermediaries and Violence Against Women Online*, Association for Progressive Communications (APC), July, available at www.genderit.org/sites/default/upload/flow-cnyst-summary-formatted.pdf (accessed 28 January 2016).

Robinson, PH (2015), 'The moral vigilante and her cousins in the shadows', *Faculty Scholarship*, Paper 506, available at http://scholarship.law.upenn.edu/faculty_scholarship/506 (accessed 20 March 2016).

Ronson, J (2015), *So You've Been Publicly Shamed*, Picador, London.

Sandoval, G (2013), 'The end of kindness: Weev and the cult of the angry young man', *The Verge*, 12 September, available at www.theverge.com/2013/9/12/4693710/the-end-of-kindness-weev-and-the-cult-of-the-angry-young-man (accessed 22 December 2015).

Sehgal, P (2016), 'The forced heroism of the "survivor" ', *The New York Times Magazine*, 3 May, www.nytimes.com/2016/05/08/magazine/the-forced-heroism-of-the-survivor.html?action=click&contentCollection=magazine&region=rank&module=package&version=highlights&contentPlacement=1&pgtype=sectionfront&_r=0 (accessed 21 September 2016).

Selvaratnam, N (2015), ' "I had no idea he was filming me": shocking trend sees women raped, blackmailed with footage', *SBS*, 19 February, available at (accessed 16 July 2016).

Sen, A (2012), 'Women's vigilantism in India: a case study of the Pink Sari Gang', *Online Encyclopedia of Mass Violence*, December, available at www.massviolence.org/women-s-vigilantism-in-india-a-case-study-of-the-pink-sari (accessed 15 July 2016).

Spicer, T (2014), 'I don't want my kids sitting next to a man on a plane', *Traveller*, 27 April, available at www.traveller.com.au/i-dont-want-my-kids-sitting-next-to-a-man-on-a-plane-375z6 (accessed 15 July 2016).

Tay, M (2016), 'Five women who fight the trolls', *The Weekly Review*, 21 June, available at www.theweeklyreview.com.au/meet/five-women-who-fight-the-trolls/ (accessed 15 July 2016).

United Nations Broadband Commission for Digital Development Working Group on Broadband and Gender (2015), *Cyber Violence Against Women and Girls: A World-Wide Wake-Up Call*, available at www.unwomen.org/~/media/headquarters/attachments/sections/library/publications/2015/cyber_violence_gender%20report.pdf (accessed 14 December 2015).

Valenti, J (2016), 'Insults and rape threats: writers shouldn't have to deal with this', *The Guardian*, 15 April, available at www.theguardian.com/commentisfree/2016/apr/14/insults-rape-threats-writers-online-harassment (accessed 15 April 2016).

White, A and Rastogi, S (2014), 'Justice by any means necessary: vigilantism among Indian women', in S. Ponzanesi (ed.), *Gender, Globalization, and Violence: Postcolonial Conflict Zones*, Routledge, New York and Oxon, pp. 214–28.

Wittes, B, Poplin, C, Jurecic, Q and Spera, C (2016), 'Sextortion: cybersecurity, teenagers, and remote sexual assault', *Brookings Institution*, May, available at www.brookings.edu/research/reports2/2016/05/sextortion-wittes-poplin-jurecic-spera (accessed 12 May 2016).

World Wide Web Foundation, (nd), *Web Index: Report 2014–15*, available at http://thewebindex.org/wp-content/uploads/2014/12/Web_Index_24pp_November2014.pdf (accessed 29 June 2016).

4 Internet intermediaries and online gender-based violence

Elena Pavan

Introduction: shifting the research focus on intermediaries and their responsibilities

In a research landscape that is increasingly concerned with understanding the implications of communication technologies for our societies, a thorough comprehension of the nexus between information and communication technologies (henceforth, ICTs) and gender-based violence remains far from achieved. Several factors contribute to this situation. First, the issue of online gender-based violence has long been overshadowed by a general enthusiasm for the 'positive' impacts of ICTs on political participation in particular (see, for example, Castells 2012), as well as everyday life more generally (see, for example, Rainie and Wellman 2012). Thus, only recently have observers started to engage systematically with the negative implications of the pervasiveness of ICTs – in particular, the persistence of gender inequality within the digital space (Wajcman 2010) and the ways in which communication technologies entwine with abusive behaviours directed against women and girls (Pavan 2013).

Second, even when the nexus between ICTs and gender-based violence has been considered, the research focus has been set on how ICTs facilitate already-recognised forms of harm against women – such as domestic violence, dating abuse, stalking and sexual exploitation – especially when involving protected categories such as children and adolescents (Henry and Powell 2016). In this context, extant studies consider adult victimisation less often, which has limited the development of contemporary understandings of 'digital harms' as a distinct and inherently heterogeneous category of harms that encompasses all dimensions of the 'embodied subject: the real, sensory, physical, psychical, exterior and interior lived experiences' (Henry and Powell 2015, p. 765).

The third reason for the absence of a more comprehensive account of the nexus of ICTs and gender-based violence, and which forms the focus of this chapter, is that, to date, there has been limited examination of the specific role played by ICTs and, therefore, no solid base from which to elaborate counteraction and restorative strategies that simultaneously address the social and technological facets of digital harms (Powell 2015). Only occasionally have ICTs been considered as *actors* in their own right (Pavan 2014) – *actively influencing* the ways in which digital harms are perpetrated and experienced. Initial efforts to overcome this gap in the research include recent studies that, at the crossroads between social theories emphasising the role of non-human agents (see Latour 2005) and feminist readings of power dynamics (for a review, see Padovani and Pavan 2015), engage with 'how new technologies operate, how they are gendered, how they are used, and what their impacts are' (Henry and Powell 2015, p. 773). One example, in this respect, is provided by the recent work of Massanari (2015), who investigated how the algorithms and functioning mechanisms of Reddit (a social media platform for web content rating)[1] facilitate the emergence of 'toxic technocultures' against women and their participation in the public sphere. Even more rarely, ICTs have been considered from an *organizational* perspective – that is, looking at the role played by the *corporate actors* that own, shape and regulate the communication services that form the current global ICT system. In particular, no specific attention has yet been paid to internet intermediaries – that is, corporate actors who 'give access to, host, transmit and index content, products and services originated by third parties on the Internet or provide Internet-based services to third parties' (Organisation for Economic Co-operation and Development 2010, p. 9) – such as internet service providers, web hosting services (like Godaddy.com), search engines (like Google) and participative networking platforms (like Facebook, Twitter and YouTube).

At first glance, including intermediaries among the protagonists of online gender-based violence may seem unnecessary because they are neither *directly involved* in abusive relationships nor do they have the authority to *formally sanction* abuses committed through their services. And yet their lack of direct involvement does not mean that intermediaries play no role in relation to gender-based violence. As the providers of communication services to third parties, they do have an opportunity to prevent digital harms from happening or to contribute to the restoration of justice after violations have occurred. In fact, intermediaries occupy a crucial position within the global ICT system, as they provide the organisational environment within which communication services are designed, implemented and made available to final users (Article 19 2013). However, while providing users with the necessary resources to get connected and informed, intermediaries are also entitled to determine how these resources shall be accessed and utilised through their terms of service (ToS). Thus, ToS constitute a true normative benchmark against which digital behaviours and content are deemed acceptable or not, and are therefore a key instrument for the definition, identification, prevention and punishment of online violent acts. Ultimately, although they are not visibly involved in digital harms dynamics in the same way as perpetrators, victims and formal criminal justice institutions, intermediaries can become facilitators of or, conversely, agents of positive resistance to prevention of and protection against gender-based discrimination, harassment and abuse (Shaikh 2015).

Nonetheless, the roles and responsibilities of intermediaries in relation to online gender-based violence remain unclear. Like any other business entity, intermediaries are increasingly expected not only to provide the best service possible but also to behave 'responsibly' – that is, to balance their search for profits with a commitment to 'improve the quality of life of the workforce and their families as well as of the local community and society at large' (World Business Council for Sustainable Development 2002, quoted in Soule 2009, p. 20). While it has been documented that there is a link between a socially responsible conduit and a company's reputation and profits (Soule 2009), there is no formal obligation to adopt such a stance, and there is no agreed definition of 'corporate social responsibility' (CSR). For this reason, intermediaries promote activities that align with *their* view of CSR – typically by continuously ameliorating their workplace and labour conditions, ensuring transparency at the governance level and committing to providing their users with an empowering service while ensuring their safety. For example, Twitter includes in its website a 'Public Policy' section wherein the company declares its commitment to promote 'healthy digital citizenship and online safety [by] supporting organizations that tackle online safety issues such as bullying, abuse and hate speech', as well as to empower 'women and unrepresented minorities in technology' (Twitter 2016a). But how much is this 'responsible' approach genuinely gender aware? How much of intermediaries' commitment to protect and empower their users entails a serious engagement with critical issues like online gender-based violence? Ultimately, what importance do intermediaries place on sustaining or combating such violence?

This chapter explores the answers to these questions by lifting intermediaries out of their usual invisibility and bringing them to the forefront of the discussion on the nexus between ICTs and gendered violence. In doing so, the aim is to progress current efforts to unveil the more *invisible mechanisms* that prelude and support online gender-based violence through the adoption of an organisational perspective to frame the role of ICTs within this domain. In what follows, I examine intermediaries' roles and responsibilities in relation to gender-based digital harms from three interrelated perspectives. First, I analyse intermediaries as *hidden influentials* that mostly facilitate digital harms through their ToS. Second, I look at how the absence of a systematic normative framework in the field of online gender-based violence allows intermediaries to behave as *free agents* – that is, as a special category of actors who, overall, enjoy impunity, as their responsibilities are seldom recognised. Finally, I discuss how intermediaries are also *targets* of civil society initiatives that denounce their negligence and urge them to act more responsibly and effectively to combat digital harms against women.

Intermediaries as *hidden influentials*

As discussed, intermediaries are not neutral actors because, through their ToS, they determine the benchmarks against which the acceptability of digital behaviours and content – hence, the existence of digital harms – can be assessed. Therefore, intermediaries are what observers of contemporary power dynamics have called 'hidden influentials' (see Gonzaléz-Bailón, Borge-Holthoefer and Moreno 2013) – actors who are able to significantly impact a process while remaining invisible. Thus, a first perspective from which it is possible to evaluate both the role and the responsibility of intermediaries in relation to gender-based digital harms is to examine the source of their hidden influence – the contents of their ToS.

While research efforts on this topic remain rare, one notable exception is provided by the study 'From Impunity to Justice', conducted between 2013 and 2014 by the Association for Progressive Communications (APC), an international organisation active since the 1990s in promoting the democratic and empowering use of ICTs. The study documented the multifaceted nature of online digital harms and assessed the adequacy of corporate policies and procedures put in place by intermediaries to address online gender-based violence. This project examined case studies of online violence in seven different countries – Bosnia and Herzegovina, Colombia, the Democratic Republic of Congo, Kenya, Mexico, Pakistan and the Philippines. In each case study, intermediaries' corporate user policies were examined in search of the specific provisions that relate to and can be used in cases of digital harms.[2] While the study is not representative of all possible occurrences of online gender-based violence and makes no claim to provide an in-depth comparative analysis of intermediaries' policies, it constitutes a useful entry point to discuss the main aspects of ToS that matter to online gender-based violence dynamics.

Overall, the APC study suggests that intermediaries do not use their hidden influence to create a safe and empowering environment for women. Quite the opposite, their ToS allow a multiplicity of digital harms to be realised, often under the 'veil of free speech' (Nyst 2013a). In the first place, what is generally missing from ToS is a clear definition of acts that can be considered online gendered violence. In this regard, the report highlights that some companies limit themselves to preventing users from employing their services to commit unlawful, illegal, fraudulent or abusive behaviours, while others provide their own definition of obscene and/or pornographic content as well as of hate speech or incitements to violence (Athar 2015). However, these two criteria seem limited at best. We know that the law is limited in relation to gendered violence and harm that occurs online or is facilitated by technology. As critical observers have noted, current legal frameworks lack a systematic and encompassing conceptualisation of digital harm, which means that the law is simply 'ill-equipped' to deal effectively and efficiently with this type of abuse (Henry and Powell 2015, p. 772). Moreover, issues arise also when intermediaries determine what is acceptable and what is not, in terms of use, content and their own response. It has been noted that when intermediaries lean on their own policies to regulate online content, they often act 'outside the scope of any internationally-recognized legitimate limitations on freedom of expression' (Article 19 2013, p. 3), and therefore often end up 'censoring consensual sexual expression [while] contributing to a conservative mind-set that reduces women's bodies to sex objects' (Athar 2015, p. 21).

One highly illustrative example is the controversial definition of 'nudity' provided in Facebook's Community Standards (CSs). As the topic is contained in the section on 'Encouraging respectful behaviour', it is in view of protecting and respecting the 'sensibility of some audiences . . . particularly because of their cultural background or age' that Facebook actively engages in removing pictures of female breasts that include nipples, unless shown in the context of breastfeeding or a post-mastectomy situation (Facebook 2016), while no mention is made of naked male chests or nipples (Athar 2015, p. 21). For the same reasons, Facebook does not allow photos of genitals or 'fully exposed buttocks' (Facebook 2016). However, a specific formulation like this one paradoxically ends up being too general to serve as a tool for identifying online abuses, as it is, for example, in the case of publicly accessible pages or groups with thousands of members commenting on so-called 'creepshots' – pictures of women's bodies photographed or filmed without their knowledge or consent and posted with an explicit voyeuristic aim.

Two further aspects of ToS are particularly relevant in relation to online gender-based violence: the definitions of hate speech and 'private' information. With regard to the former, observers argue that the explicit banning of aggressive and/or discriminating behaviours may have limited 'overt hate speech' but has not proved efficient in containing 'covert discrimination' that takes place – for example, through stigmatisation, the systematic use of slurs or the sharing of discriminatory images (Ben-David and Matamoros-Fernández 2016, p. 1168). In this sense, platforms such as Facebook and Twitter draw a blurred line between illicit hate speech acts and acceptable 'humorous', 'satiric' or 'social commentary' acts and thereby allow the circulation of misogynistic messages or even slurs. For example, in the summer of 2013, a group of Italian Facebook users who joined the national initiative against femicide '*Ferite a Morte*' ('Wounded to Death') asked the platform to remove the advertisement of a restaurant they perceived as inciting violence against women. The restaurant promoted the quality of its food with the slogan '*Solo carne fresca*' ('Only fresh meat') alongside graphic content picturing a chopped and bleeding female body in a bikini. While the picture was repeatedly reported as graphic sexual violence, Facebook did not remove it because it did not violate its CSs, which define 'sexual violence and exploitation' with specific reference to 'sexual exploitation of minors and sexual assault' (Facebook 2013).

When it comes to user privacy, intermediaries tend to set very loose boundaries to distinguish between private and public content and thus to protect women's safety. For example, Facebook leaves its users to define levels of content publicity by making use of the privacy setting.[3] However, some content, like cover and profile images, is public by default. In this sense, commentators underline that the loose definition of 'private' information, in combination with the 'real name policy' pursued by the company, can threaten activists' safety (Hintz 2016). For instance, the APC documents the case of a Pakistani women's rights defender whose Facebook pictures were stolen and reused in posters that characterised her and her campaigning efforts as blasphemous. As a consequence of the diffusion of these posters, the activist suffered increasing threats and harassment both online and offline, to the point that she opted for self-censorship, suspending her and her family members' social media accounts (GenderIT 2014).

Conversely, Twitter makes a list of 'private and confidential information' that users should refrain from posting or whose publication can be opposed (such as intimate photos or videos that were taken or distributed without the subject's consent), but the platform's policy also states that 'not all posting of such information may be a violation of this policy. We may consider the context and nature of the information posted, local privacy laws, and other case-specific facts when determining if this policy has been violated' (Twitter 2016b). In particular, Twitter specifies that a piece of content 'previously posted or displayed elsewhere on the Internet prior to being put on Twitter' may not violate the platform's privacy policy (Twitter 2016b). Analogously, Reddit's policy states that 'we are not responsible for the content or actions of any third party websites or services associated with posted links' (Reddit 2016). On the one hand, this formulation functions to avoid charging a specific intermediary for a violation committed through another online service. On the other, it leaves plenty of room for reiterating abuses without necessarily taking responsibility for doing so. For example, Massanari (2015) discusses two cases – 'the Fappening' and 'Gamergate' – in which private information about two female celebrities (actress Jennifer Lawrence and the game developer Zoe Quinn, respectively) was broadly circulated and gained high visibility on Reddit. In both cases, the platform's administrators did not ban or remove any harmful, offensive or threatening content or comments targeted at the two victims under the justification that 'the platform was merely distributing the content rather than *actually hosting it*' (2015, pp. 11–12, emphasis added).

Strictly connected to the issue of privacy is that of individual consent to the circulation and use of data and information, which becomes particularly relevant in relation to 'nonconsensual pornography – i.e., the distribution of sexually graphic images of individuals without their consent' (Citron and Franks 2014, p. 346). Looking, in particular, at the policies of pornography websites like XVideos and YouPorn, the APC study notes that intermediaries often include in their ToS a clause that binds the possibility to publish content to the possession of third-party consent. However, users who upload pictures or videos on this type of platform are not requested to prove their possession of such consent (Athar 2015, p. 33). For this reason, materials retrieved from other platforms can be manipulated and reposted, thus making women involuntary protagonists of pornographic videos (GenderIt 2015); or the non-consensual publication of intimate pictures or shots can be used by exes as 'revenge porn', causing severe damage to the victims, equal to that caused by domestic violence and sexual assault (Citron and Franks 2014). In not tuning their ToS in a way that allows them to either tackle or sanction such occurrences as actual forms of abuse, intermediaries are in fact exerting a hidden influence in shaping an online space of interaction that is not only tolerant towards but, more relevantly, reinforces discriminatory and abusive behaviours against women.

Intermediaries as *free agents*

The review of company policies performed by the APC clearly shows that 'the standards across the ToS of many internet intermediaries are primarily reflections of their legal obligations, and not much more than that' (Athar 2015, p. 35). Because of the absence of a systematic and effective normative framework in the field of online violence at both the national and international levels (Citron and Franks 2014; Henry and Powell 2015; Powell 2015), intermediaries have no particular obligation to restrict their ToS in view of either preventing or sanctioning gender-based digital harms. Moreover, because there is no normative benchmark against which the responsiveness of their ToS can be measured, intermediaries' responsibilities in cases of online violence are seldom recognised. Within this normative void, intermediaries ultimately can behave as *free agents* – that is, as a special category of actor who is not subjected to any particular obligation and who, ultimately, enjoys overall impunity when women's rights violations occur.

Indeed, over the past decade, the subject of online gendered violence has found some space within the official and ever-evolving supranational discourse on violence against women and girls. In particular, the 2006 *In-Depth Study on All Forms of Violence against Women* report called for more attention on 'the use of technology in developing and expanding forms of violence' (UNGA 2006, para 155). More recently, while examining the situation in Great Britain and Northern Ireland, the UN Special Rapporteur on Violence Against Women, Rashida Manjoo, recognised 'internet violence' as a form of abuse and associated it with 'harmful behaviours online, including humiliation, harassment, intimidation and "sexting" as a form of bullying' as well as with the actual or threatened publication of 'consensual sexual activity to harass and manipulate women and even force them to engage in sexual activity with the boyfriend and/or their friends' (UNGA 2015, para 25).

A more explicit attempt to define the issue was made by the Broadband Commission for Digital Development (BCDD) in 2015 in its report *Cyber Violence against Women and Girls: A World-wide Wake-up Call.* According to the report, 'Cyber VAWG [violence against women and girls] includes hate speech (publishing a blasphemous libel), hacking (intercepting private communications), identity theft, online stalking (criminal harassment) and uttering threats. It can entail convincing a target to end their lives (counselling suicide or advocating genocide)' (BCDD 2015, p. 6). Thus, the report identifies six different categories of online abusive behaviour: hacking, impersonation, surveillance/tracking, harassment/spamming, recruitment and malicious distribution of defamatory/illegal materials (BCDD 2015, p. 22). To counteract these behaviours, the report calls for a multi-level approach that passes through three main steps: *sensitisation*, in order to change the societal norms that routinise gender inequalities; *safeguarding* – that is, the continuous collaboration with intermediaries and users to guarantee the online safety of all users, in particular women and girls; and *sanctioning*, which points to the construction and implementation of adequate legal frameworks to 'enforce compliance and punitive consequences for perpetrators' (BCDD 2015, p. 3).

Soon after its release, the BCDD report was heavily criticised and was eventually withdrawn by the Commission itself on October 2015, pending review. Critical observers identified two main reasons that yielded to the report's withdrawal. First, the report was deemed to be drafted using 'poor materials' and to contain several errors, especially within the footnotes (Järvinen 2015). Second, the formal inaccuracy of the report was connected to a substantial inconsistency in its proposals, which some observers claimed were an attempt of the United Nations (UN) to legitimise 'internet censorship' (Mikelionis 2015). At the crossroads between formal and substantial matters, the report withdrawal transformed the policy discussion around online gendered violence into an actual controversy in which the very concept of online gender-based violence was questioned and publicly portrayed as instrumental to governmental manoeuvres to repress dissent and violate free speech (White 2015). Certainly, as some commentators have argued, the BCDD report withdrawal was a 'missed opportunity' (Jeong 2015) to recognize the actual existence and gravity of online gender-based violence as well as to overcome the normative void around it. However, a closer look at the issues that fuelled this controversy can help us identify the true challenges that make the definition of a normative framework for online gender-based violence particularly difficult.

The complex matter of intermediaries' liability – that is, their legal responsibility in relation to illegal or harmful content authored by third parties and circulated through their services – represents one such challenge. Because intermediaries cannot control all the content produced, exchanged or simply referenced on the platforms they provide, international standards grant them immunity when violations occur unless they fail to comply with specific requirements, typically 'removing or disabling access to "illegal" information when they obtain actual knowledge of such content' (Article 19 2013, p. 7). Thus, intermediaries' non-liability is traditionally seen as one of the cornerstones of internet freedom as well as a strategic element in the continuous improvement of the global communication system (Nyst 2013b).

Requests to remove or block access to digital content are often issued in relation to copyright infringements or other legal issues – such as privacy violations, child protection or defamation. For example, the Web Index Report, the principal measure of how the web contributes to social, economic and political progress in 86 countries,[4] states that, in the first half of 2013 alone, Google received copyright takedown notices for about 4 million URLs per week and governmental removal requests for about 1000 items per week (World Wide Web Foundation 2014, p. 13). While recognising that the majority of these requests were indeed 'directed towards piracy and other legitimate instances of infringement', the report affirms that 64 per cent of surveyed countries 'have not established clear and adequate protection for intermediaries' and that, given such legal uncertainty, intellectual property protections are 'being abused to deny fair use, disadvantage competitors, or suppress criticism' (World Wide Web Foundation 2014, p. 13). Moreover, the report emphasises that online gendered violence remains a poorly addressed issue, as 74 per cent of the surveyed countries, 'including many high-income nations, law enforcement agencies and the courts are failing to take appropriate actions in situations where web-enabled ICTs are used to commit acts of gender-based violence' (World Wide Web Foundation 2014, p. 15).

While intermediaries' involvement in online gendered violence continues to be discussed only in relation to their formal liabilities rather than framed in terms of their 'social responsibility', proposals that entail an increased level of control of online content, like those advanced by the 'Safeguards' section of the BCDD report, are immediately seen by many as connected to a 'restrictive approach that endangers the free and open nature of the internet and implies a risk-based consideration' (Nyst 2013b, para 4). In this sense, the issue of online gender-based violence is considered as a shield to justify censorship and thereby, paradoxically, becomes a threat to other human rights. For this reason, it gets quickly discarded in favour of further-reaching claims of free-speech defence – albeit forgetting that the international human rights framework does not conceive of freedom of expression as an absolute right and foresees limitations, particularly in order to respect others' rights and reputation and avoid justifying hatred and discrimination (Nyst 2013b).

Even more problematic is the fact that critics who oppose the BCDD work and dismiss online gendered violence base their arguments on the assumption that 'harassment online does not amount to violence unless there is the probability of "imminent harm" or "real violence" ' (Nyst 2013c, point 2). For example, as stated in a commentary to the BCDD report, 'even the stupidest voice on the internet is just making empty noise. If and when it morphs from noise into real life conduct, it becomes a different problem. *Then* it's real. *Then*, it can be actionable violence. But as long as it's virtual, *it's just the noise* made by the idiot children' (Greenfield 2015, para 13, emphasis added). Objections of this sort in fact endorse what Nyst calls 'key gendered norms and assumptions': first, that violence, in order to be defined as such and to be actable upon, must be physical; second, that 'speech that trivializes and glorifies violence against women does not amount to hate speech'; and, third, that 'common misogynistic slurs do not present a real threat of violence' (2013c, points 1–4).

In turn, these assumptions crystallise 'invisible power relations between genders' (APC, JASS and Women's Net 2015, p. 27) that, precisely by virtue of their invisibility, are inevitably reproduced in all domains, even in the formulation of universal principles like human rights. Indeed, in reviewing the history of struggles for women's communication rights, Gallagher underlines that 'the abstract, genderless, rights-bearing individual posited in "universal" rights documents and declarations is, in fact, conceived – or "embodied" – as a male political subject' (2014, p. 235). The underlying masculine perspective that permeates global normative frameworks, coupled with the impossibility of clearly affirming a gender perspective within key institutional processes (just to name one, the UN World Summit on the Information Society), has continued to push the specific needs of women and girls to the margins, thus conveying prominence to more far-reaching and (allegedly) ungendered arguments, like the need to guarantee a human right of free speech or intellectual property.

In this context, intermediaries are not compelled to deal with reports of online gender-based abuse in any particular way and thus can be understood to work as free agents who operate with limited accountability. The APC report finds that intermediaries are indeed poorly transparent and accountable when it comes to explaining their lack of action in response to users' requests for intervention through the available reporting procedures or tools (Athar 2015, p. 26). While automated messages can be sent to communicate that flagged content does not violate the platform ToS, users often have no idea of how their reports are handled, if handled at all; of who is in charge of evaluating them; or of the level of training the staff has in relation to the recognition and the counteraction of digital harms (Athar 2015). Moreover, the APC study identifies language as a major obstacle to enjoying the right to redress: reporting systems are often available only in English, which prevents many users from being able to reach the intermediary through the established channels while having no true alternative way to convey their remarks.

In parallel to the APC study, the organisation Women, Action and the Media (WAM!) analysed how Twitter handles harassment reports submitted by its users (Mathias et al. 2015). As a Twitter 'authorized reporter',[5] WAM! collected and transmitted genuine harassment reports made by either victims or bystanders of abuse that occurred upon the microblog platform – from doxing (the release of private information without consent), hate speech and threats of violence to impersonation and the publication of false information. The organisation found that Twitter took action in the case of 55 per cent of reports escalated, particularly (but not only) when WAM! reporters underlined the risks posed to the victims' personal safety. Thus, the platform took action against perpetrators more often than it declined WAM!'s request for action, principally by suspending the harasser's account. Nonetheless, the study also found several problematic aspects in Twitter's reporting mechanisms. Indeed, 18 per cent of users claimed to have notified the platform more than once of repeated abuse, signalling the difficulties of seeking to denounce sustained harassment. Also, major difficulties were encountered by first-time receivers of abuse, in particular, in complying with Twitter's standard reporting procedure. Twitter does not accept screenshots of tweets as evidence; only URLs can be supplied to corroborate a request for action. However, practices like doxing are often accompanied by precautionary measures performed by abusers, like the so-called 'tweet and delete' practice by which unauthorised content is published and then quickly removed before reparatory action can be taken (Mathias et al. 2015, p. 23). This action means that, from the time of the tweet appearing to the reporting of the incident in line with the procedures set out by Twitter, the abusive content may be already cancelled, so the victim is no longer able to have the harm of the short-term message/image recognised.

In the absence of any formal pressure to respond to abusers working around ToS and content rules, intermediaries are not obliged to take any particular preventive or restorative course of action in relation to digital harms. Quite the opposite; their status of free agent is defended and protected as a precondition to the exercise of human rights. Ultimately, when it comes to cases of gender-based violence, the overall immunity status intermediaries enjoy often turns into overall impunity, to the detriment of women's rights and safety.

Intermediaries as *targets* of civil society initiatives

The complex situation outlined in the previous section suggests that a clearer legal framework to address digital harms is needed as much as it remains necessary to clearly articulate intermediaries' formal obligations in this domain. However, as poignantly noted by Powell (2015), confining the possibility of combating and defeating online gendered violence to the identification of a legal framework implies binding the very concept of 'justice' to 'punitive state-sanctioned outcomes [thus] continuing to elude the vast majority of rape victim-survivors' (2015, p. 573). In this sense, emphasising intermediaries' responsibility not as a substitution for but as a necessary complement to their formal liabilities seems a necessary step towards a more comprehensive approach to preventing and tackling digital harms against women.

International organisations do in fact recognise that intermediaries bear some responsibility for the protection of human rights. For example, the Recommendations of the Organization for Security and Co-operation in Europe (OSCE) Representative on Freedom of the Media frame the online abuse of female journalists as a 'new challenge' to freedom of expression and recognise that intermediaries play a role in relation to it. Starting from the premises that 'journalists' safety is a precondition to free speech and free media' and that the safeguard of female journalists' human rights and freedom of expression must be ensured (OSCE 2015, pp. 5–6), the recommendations call intermediaries to 'ensure that *terms of service, community guidelines* and *information about their enforcement* are *proportionate* and *adequate*, clear, understandable and easily available to all users' (OSCE 2015, p. 7, emphasis added).

However, a much stronger call to intermediaries to act responsively has come from the heterogeneous realm of civil society. At the crossroads between a long tradition of social movements targeting corporate actors (Soule 2009) and the increasing relevance of online forms of feminist activism and resistance (Powell 2015), a transnational network of civil society and grassroots organisations, activists, journalists, bloggers, academics and citizens is coordinating to reclaim ICTs to end all forms of violence and solicit all stakeholders, in particular intermediaries, to take on their responsibilities in counteracting online gendered violence. Just as non-state actors have played important roles in providing victims and survivors of violence with 'alternative and innovative justice mechanisms and community-led practices' to restore justice (Powell 2015, p. 581), these civil society initiatives have been crucial in unveiling intermediaries from their overall invisibility and exposing their practices in relation to digital harms to public scrutiny.

There are several ways in which civil society organisations and activists are turning intermediaries into *targets* of public attention in relation to online gendered violence. First and foremost, activists engage in 'corporate campaigns' that attack intermediaries' public image 'with the hope of getting the company to agree to a set of demands' (Soule 2009, p. 13) – in this case, a more proactive stand against gendered violence. Two examples are particularly worth mentioning in this regard. The first is the #FBrape campaign promoted in 2013 by the Everyday Sexism Project and WAM!. Making systematic use of the hashtag #FBrape, activists circulated more than 60,000 tweets bearing evidence of the many violent uses made of Facebook. As a result, major brands (including Nissan UK) pulled their advertising from the platform, thus compelling the social network to take a more systematic approach towards the moderation of violent content (Buni and Chemaly 2014). Buried by the evidence of their 'irresponsible' behaviour and concerned with the economic and reputational losses deriving from the campaign, Facebook issued a press release in which the company admitted 'that our systems to identify and remove hate speech have failed to work as effectively as we would like, particularly around issues of gender-based hate. . . . We need to do better – and we will' (Facebook 2013).

A second example is the multilingual campaign launched in 2014 by the APC with the hashtag #whatareyoudoingaboutvaw as a spinoff of its annual campaign on online gendered violence Take Back The Tech! (Pavan 2015). Appearing less than a year from the #FBrape campaign, the circulation of #whatareyoudoingaboutvaw revamped attention towards intermediaries' responsibilities, thus generating a worldwide discussion on traditional and digital media and fostering multiple partnerships, such as that with UNiTE – the UN Secretary-General's campaign to End Violence Against Women (Fascendini 2014).

The combination of corporate campaign and 'advocacy science' (Soule 2009, p. 15) – that is, the production of alternative content and knowledge (Fuster Morell 2009) – is another way in which companies are being pressured from below. Reports like those produced by the APC and WAM! have provided alternative and unprecedented analyses of intermediaries' normative and procedural standards in relation to online gendered violence and have thus pressured them to act more responsively. For example, in conjunction with the publication of the WAM! report, Twitter has allowed bystanders to report cases of online abuse they witness on the platform and has begun to provide users who report violations with an email record to be shared with law enforcement agencies (Mathias et al. 2015, p. 31).

Besides targeting intermediaries directly through campaigns and advocacy science, activists and civil society organisations are seeking to resolve the conflict between online services companies' formal liabilities and their more informal, but no less important, responsibilities in relation to online gendered violence. In order to do so, activists are pursuing a systematic engagement with institutions and, in this way, contributing to broader political processes with their strategic thinking (Fuster Morell 2009, p. 23). Exemplary here is the systematic participation of organisations like the APC in the Internet Governance Forum (IGF), the official UN process to discuss current and future development in the governance of the internet system; and the Commission on the Status of Women (CSW), one of the principal supranational institutional forums discussing of issues of gender equity and women's rights. In the context of the IGF, the APC has joined efforts with other organisations and initiatives, thus forming the Dynamic Coalition on Gender and Internet Governance (DCGIG), a group that aims to ensure gender parity and inclusion within the IGF process – a necessary condition to push issues such as online gendered violence and intermediaries' responsibilities towards it within the international agenda (DCGIG 2015). Within the CSW context, instead, activists are organising several initiatives to address the issue of online gendered violence and to urge a systematic reflection on the role played by intermediaries in this domain. During the 2015 edition of the CSW, the APC discussed the results of its report on intermediaries' policies and practices during an ad hoc panel in which participants proposed a set of alternative solutions through which intermediaries could proactively combat online gendered violence. In particular, activists suggested that existing mechanisms be employed, such as the *Convention on the Elimination of All Forms of Discrimination against Women* – a binding provision representing a general 'bill for rights for women' – which could serve as a starting point to calibrate possible exceptions to the universal right to freedom of expression (Fascendini 2015).

Conclusion

Over the past few years, especially in response to organised campaigns and, to be fair, partly as a result of lawsuits issued by victims, intermediaries have adopted a more responsible approach towards online gendered violence. This has sometimes involved publicly responding to claims raised by organisations and activists and admitting their inability to take on their responsibilities and sometimes revising their ToS to reduce, at least in principle, the risk of abuse (Buni and Chemaly 2014). Practical and governance measures have also been implemented: Facebook has set up a Safety Advisory Board comprised of five European and North American organisations; Twitter collaborates with expert organisations appointed as 'authorised reporters'; and Instagram has recently released a feature that allows users to filter out offensive comments.

However, much remains to be done in this area, and necessary steps must be taken. A first step consists of *recognising the socio-technical nature of online gendered violence*. Acts of online abuse and harassment are a complex and ever-evolving combination of technological affordances and socio-cultural practices. It is therefore necessary to start precisely from the socio-technical practices and not from some aprioristic assumption on what online gendered violence *is* or *should be*, in order that such complex and delicate matters can be tackled. What is crucial here is recognising the knowledge produced by movements and organisations as *legitimate* and *necessary*. Indeed, intermediaries have so far drafted and applied their ToS following what Gramsci (1991, in Cox 2014) has called a 'common sense knowledge' – one that is imbued with hegemonic social relationships and crystallises inequalities. To find workable solutions and follow useful suggestions to take up their responsibilities in defending women's rights, intermediaries can lean on what Gramsci terms 'good sense knowledge', nurtured by popular practices like those deployed by organisations and activists mobilising in what could be called an 'ICT for gender movement'. An understanding of the implications of adopting a Gramscian view of this approach warrants further research and development elsewhere, but this chapter has laid the groundwork for such exploration.

Further, it is critical to recognise that intermediaries necessarily hold both formal and informal responsibilities. Developing and implementing more gender-aware ToS and improving the current mechanisms for dealing with cases of online gender-based violence are essential first steps. Far from requiring a complete redraft of privacy and security settings or content reporting procedures, the inclusion of movement knowledge within intermediaries' procedures can more easily be achieved by including female engineers, programmers and managers; by inviting experts to the company's table; and by creating a gender-aware and balanced work system that is also receptive of non-western approaches to gender (Athar 2015; Buni and Chemaly 2014; Nyst 2013c). Also, the inclusion of a gender-based perspective is fundamental when it comes to designing the algorithms and mechanisms that *deliver* reports to companies' departments and staff members involved in redressing violations (Crawford and Gillespie 2016). As much as technology is not neutral when it comes to the possibility of *committing* abuses, it remains a non-neutral factor when remedying processes are set in motion.

Recognising and handling the socio-technical nature of online gendered violence entails, in turn, acknowledging that *content reporting or flagging systems are much more than a tool* offered to users to participate in the curation and governance of a communication platform. As Crawford and Gillespie note, 'while a flag may appear to be a single data point – something is reported or it is not – this simplicity belies a tangle of system designs, multiple actors and intentions, assertions and emotions' (2016, p. 411). While much more work must be done to make users' participation truly universal (for example, by making reporting systems available in more languages), more transparency should be ensured in how submitted reports are dealt with and more systematic collaboration procedures should be put in place so that intermediaries can team up with law enforcement agencies. In this sense, intermediaries' employees in charge of dealing with reports must be effectively trained to recognise and deal with the complexities of online gendered violence so that they are able to detect online harassment and violence and decide on the right course of action – from the substitution of a mobile telephone number for mobile operators to content removal and account suspension for social media platforms and escalation to law and enforcement agencies.

Last but not least, an important step consists in making CSR genuinely gender aware by investing resources and time towards the systematic and sustained protection and empowerment of women. At first sight, acting to resize the boundaries of online interactions and contents in view of protecting women's safety and fostering their freedom of expression may not appear a profitable business. Indeed, in some cases online abuses of women have generated a significant amount of online traffic and hence of revenue – as Massanari documents in discussing the cases of the Fappening and Gamergate. This notwithstanding, enlarging the scope of CSR to genuinely include gender issues is a task that, crucially, important actors like intermediaries must undertake in order to make our society not only global but, more importantly, truly free and fair.

Notes

1 Reddit can be found at www.reddit.com.

2 More information on the methodology developed by the project to assess the responsiveness of legal and corporate policy frameworks as well as on how single case studies were collected and analysed can be found at www.genderit.org/onlinevaw/.

3 Facebook's Statement of Rights and Responsibilities affirms that 'When you publish content or information using the Public setting, it means that you are allowing everyone, including people off of Facebook, to access and use that information, and to associate it with you (i.e., your name and profile picture)' (point 2[4]).
4 Source: http://thewebindex.org/about/#how_were_countries_chosen_for_inclusion_in_the_web_index?
5 Authorized reporters are officially allowed to submit to the platform reports on behalf of other users.

References

APC, JASS and Women's Net (2015), *ICTs' for Feminist Movement Building: Activist Toolkit*, online, available at www.apc.org/en/system/files/ICTs%20Toolkit_2015_0.pdf (accessed 7 July 2016).

Article 19 (2013), *Internet Intermediaries: Dilemma of Liability*, available at www.article19.org/data/files/medialibrary/37242/Intermediaries-BAHASA.pdf (accessed 7 July 2016).

Athar, R (2015), *From Impunity to Justice: Improving Corporate Policies to End Technology-Related Violence Against Women*, available at www.genderit.org/sites/default/upload/flow_corporate_policies_formatted_final.pdf (accessed 7 July 2016).

Ben-David, A and Matamoros-Fernández, A (2016), 'Hate speech and covert discrimination on social media: monitoring the Facebook pages of extreme-right political parties in Spain', *International Journal of Communication*, vol. 10, pp. 1167–93.

Broadband Commission for Digital Development [BCDD] (2015), *Cyber Violence Against Women and Girls: A World-Wide Wake-Up Call*, United Nations Entity for Gender Equality and the Empowerment of Women and United Nations Development Programme (UNDP) Publication.

Buni, C and Chemaly, S (2014), 'The unsafety net: how social media turned against women', *The Atlantic*, available at www.theatlantic.com/technology/archive/2014/10/the-unsafety-net-how-social-media-turned-against-women/381261/ (accessed 8 July 2016).

Castells, M (2012), *Networks of Outrage and Hope*, Polity Press, Cambridge.

Citron, DK and Franks, MA (2014), 'Criminalizing revenge porn', *Wake Forest Law Review*, vol. 49, pp. 395–1.

Cox, L (2014), 'Movements making knowledge: a new wave of inspiration for sociology?', *Sociology*, vol. 48, no. 5, pp. 954–71.

Crawford, K and Gillespie, T (2016), 'What is a flag for? Social media reporting tools and the vocabulary of complaint', *New Media and Society*, vol. 18, no. 3, pp. 410–28.

Dynamic Coalition on Gender and Internet Governance [DCGIG] (2015), *Recommendations*, available at www.intgovforum.org/cms/documents/igf-meeting/igf-2016/826-dcgig-2015-output-document-1/file (accessed 7 July 2016).

Facebook (2013), *Controversial, Harmful and Hateful Speech on Facebook*, available at www.facebook.com/notes/facebook-safety/controversial-harmful-and-hateful-speech-on-facebook/574430655911054 (accessed 28 September 2016).

Facebook (2016), *Community Standards*, available at www.facebook.com/communitystandards.

Fascendini, F (2014), *Take Back the Tech! #whatareyoudoingaboutVAW: Media Report*, available at www.apc.org/fr/system/files/APC_TBTT_Intermediaries_MediaReport_September2014.pdf (accessed 7 July 2016).

Fascendini, F (2015), *Section J: From Footnotes to Headlines*, Association for Progressive Communications, available at www.apc.org/en/node/20266 (accessed 7 July 2016).

Fuster Morell, M (2009), 'Action research: mapping the nexus of research and political action', *Interface*, vol. 1, no. 1, pp. 21–45.

Gallagher, M (2014), 'Reframing communication rights: why gender matters', in C. Padovani and A. Calabrese (eds), *Communication Rights and Social Justice: Historical Accounts of Transnational Mobilizations*, Palgrave Macmillan, London, pp. 234–248.

GenderIT (2014), *When a Morphed Video Leads to Further Harassment*, available at www.genderit.org/sites/default/upload/case_studies_mex3_1.pdf (accessed 7 July 2016).

GenderIT (2015), *When Women's Human Rights Are Deemed 'Blasphemous'*, available at www.genderit.org/sites/default/upload/case_studies_pak1_1.pdf (accessed 7 July 2016).

González-Bailón, S, Borge-Holthoefer, J and Moreno, Y (2013), 'Broadcasters and hidden influentials in online protest diffusion', *American Behavioral Scientist*, vol. 57, no. 7, pp. 943–65.

Greenfield, SH (2015), 'The new virtual violence', *Simple Justice: A Criminal Defense* Blog, available at http://blog.simplejustice.us/2015/09/26/the-new-virtual-violence/ (accessed 7 July 2016).

Henry, N and Powell, A (2015), 'Embodied harms: gender, shame, and technology-facilitated sexual violence', *Violence Against Women*, vol. 21, no. 6, pp. 758–79.

Henry, N and Powell, A (2016), 'Technology-facilitated sexual violence: a literature review of empirical research', *Trauma, Violence and Abuse*, DOI: 10.1177/1524838016650189.

Hintz, A (2016), 'Restricting digital sites of dissent: commercial social media and free expression', *Critical Discourse Studies*, vol. 13, no. 3, pp. 325–40.

Järvinen, H (2015), 'UN withdraws the report on cyber violence against women', *EDRI – European Digital Rights*, available at https://edri.org/un-withdraws-report-cyber-violence-against-women/ (accessed 7 July 2016).

Jeong, S (2015), '"I'm disappointed": Zoe Quinn speaks out on UN cyberviolence report', *Motherboard*, available at http://motherboard.vice.com/read/im-disappointed-zoe-quinn-speaks-out-on-un-cyberviolence-report (accessed 7 July 2016).

Latour, B (2005), *Reassembling the Social*, Oxford University Press, London.

Massanari, A (2015), '#Gamergate and the Fappening: how Reddit's algorithm, governance, and culture support toxic technocultures', *New Media and Society*, published online 9 October, DOI: 10.1177/1461444815608807.

Mathias, JN, Johnson, A, Boesel, WE, Keegan, B, Friedman, J and DeTar, C (2015), *Reporting, Reviewing, and Responding to Harassment on Twitter: Women, Action, and the Media*, May 13, 2015, available at http://womenactionmedia.org/twitter-report (accessed 7 July 2016).

Mikelionis, L (2015), 'The UN proposal on cyberbullying is a blueprint for censorship', *The Telegraph*, available at www.telegraph.co.uk/technology/social-media/11902140/The-UN-proposal-on-cyberbullying-is-a-blueprint-for-censorship.html (accessed 7 July 2016).

Nyst, C (2013a), 'Pulling back the veil of free speech', *GenderIT*, available at www.genderit.org/es/node/3921 (accessed 7 July 2016).

Nyst, C (2013b), 'Towards internet intermediaries' responsibility', *GenderIT*, available at www.genderit.org/feminist-talk/towards-internet-intermediary-responsibility (accessed 7 July 2016).

Nyst, C (2013c), 'How gender-based harassment falls through the digital cracks', *GenderIT*, available at www.genderit.org/es/node/3920 (accessed 7 July 2016).

Organisation for Economic Co-operation and Development [OECD] (2010), *The Economic and Social Role of Internet Intermediaries*, available at www.oecd.org/internet/ieconomy/44949023.pdf (accessed 7 July 2016).

Organization for Security and Co-operation in Europe [OSCE] (2015), *New Challenges to Freedom of Expression: Countering Online Abuse of Female Journalists*, published by the OSCE Representative on Freedom of the Media.

Padovani, C and Pavan, E (2015), 'Networks and power in gender-oriented communication governance: linking theory to empirical explorations', in N Chitty and L Qing (eds), *China, USA and the World: Theatres of Soft Power*, Zhongguo chuan mei da xue chu ban she, Beijing, pp. 42–68.

Pavan, E (2013), 'Collective action and Web 2.0: an exploratory network analysis of Twitter use during campaigns', *Sociologica*, vol. 3, pp. 1–29.

Pavan, E (2014), 'Embedding digital communications within collective action networks: a multidimensional network perspective', *Mobilization*, vol. 19, no. 4, pp. 441–55.

Pavan, E (2015), '#takebachthetech and #WhatAreYouDoingAboutVAW: reclaiming ICTs and soliciting stakeholders' responsibility to end violence against women', *Feminist Media Studies*, vol. 15, no. 1, pp. 159–62.

Powell, A (2015), 'Seeking rape justice: formal and informal responses to sexual violence through technosocial counter-publics', *Theoretical Criminology*, vol. 19, no. 4, pp. 571–88.

Rainie, L and Wellman, B (2012), *Networked: The New Social Operating System*, The MIT Press, Cambridge, MA.

Reddit (2016), *User Agreement*, available at www.reddit.com/help/useragreement/ (accessed 28 September 2016).

Shaikh, R (2015), *Violence Against Women Online: What Next Steps Intermediaries Should Take*, available at www.genderit.org/node/4216/.

Soule, S (2009), *Contention and Corporate Social Responsibility*, Cambridge University Press, Cambridge and New York.

Twitter (2016a), *Twitter for Good*, available at https://about.twitter.com/company/twitter-for-good (accessed 28 September 2016).

Twitter (2016b), *Private Information Posted on Twitter*, available at https://support.twitter.com/articles/20169991 (accessed 28 September 2016).

United Nations General Assembly (2006), *A/61/122/Add.1 – In-Depth Study on All Forms of Violence Against Women: Report of the Secretary-General*, available at https://documents-dds-ny.un.org/doc/UNDOC/GEN/N06/419/74/PDF/N0641974.pdf?OpenElement (accessed 7 July 2016).

United Nations General Assembly (2015), *A/HRC/29/27/Add.2 – Report of the Special Rapporteur on Violence Against Women, Its Causes and Consequences, Rashida Manjoo*, addendum, available at www.ohchr.org/EN/Issues/Women/SRWomen/Pages/AnnualReports.aspx (accessed 7 July 2016).

Wajcman, J (2010), 'Feminist theories of technology', *Cambridge Journal of Economics*, vol. 34, no. 1, pp. 143–52.

White, K (2015), 'Revisiting the UN broadband commission's "cyberviolence" report', *Popehat*, available at https://popehat.com/2015/09/28/revisiting-the-un-broadband-commissions-cyberviolence-report/ (accessed 7 July 2016).

World Wide Web Foundation (2014), *Web Index Report 2014–15*, available at http://thewebindex.org/wp-content/uploads/2014/12/Web_Index_24pp_November2014.pdf (accessed 7 July 2016).

5 Anti-rape narratives and masculinity in online space

A case study of two young men's responses to the Steubenville rape case

Fairleigh Gilmour and Laura Vitis

Introduction

On 11 August 2012, in Steubenville, Ohio (United States), two teenage boys, Trent Mays and Ma'lik Richmond, raped a drunk and unconscious 16-year-old girl after a night of drinking at several parties. The victim learned of her assault the next day after images and recordings of the evening were circulated on phones, and the rape was discussed on social media. Mays and Richmond were later convicted for sexual assault in June 2013, with Mays also convicted of disseminating child pornography for sharing images of the victim taken during the offence.

This chapter draws from a social constructionist understanding that emphasises the importance of rape narratives in legitimising a cultural climate in which violence against women (VAW) is normalised and endemic. Dominant western media discourse around sexual violence often constructs it as an inevitable fact of life, positioning women as perpetually vulnerable and men as naturally violent (Campbell 2005; Marcus 1992). Media analysis has demonstrated that print reportage on the Steubenville case continued the tradition of reifying rape myths by focusing on the victim's intoxication and framing the perpetrators as fallen heroes (Gutsche and Salkin 2015; Pennington and Birthisel 2016). Social media, however, played a key role in this case – it was the 'rape that social media brought to light' (Oppel 2013). Social media discussion on the case marked a shift away from the historical tropes of rape myths and aided in the distribution of 'feminist reporting on rape culture and critiques of slut shaming . . . particularly among younger feminists' (Rentschler 2014, p. 68). We argue that social media responses to Steubenville also highlighted that such contestations occur outside traditional feminist circles. To explore this observation, we undertake a case study that explores how two young men – Deric Lostutter, a hacktivist aligned with Anonymous, and Philip DeFranco, a vlogger[1] – attempted to challenge aspects of rape culture in their online responses to the case.

This case study examines how digital technology offers a platform through which to generate counter-publics (Fraser 1990) that deconstruct rape culture and catalyse discussion around gendered violence. Much of the emergent criminological analysis of online anti-rape and -harassment movements has focused on the political practice of girls and women (Jane 2012; Vitis and Gilmour 2016). In this chapter, we contribute to this burgeoning scholarship on online resistance by exploring the complexities of young men's responses to gendered violence. The first case study focuses on Deric Lostutter, a hacker whose activism influenced how the case was discussed at a national and international level. Lostutter, through his engagement with the online collective Anonymous, made a video in which he threatened to dox[2] anyone associated with the assault. From other Anonymous hackers, he solicited images, videos and information about the night which he then made public and which were pivotal in garnering public anger about the case (particularly a photo of the unconscious victim being carried by the two perpetrators and a 12-minute video in which the perpetrators' friends joked about the assault). While Anonymous's vigilante approach to social justice is popular (albeit polarising), their approach to gendered violence, including Lostutter's response to Steubenville, has been subject to feminist critique for striving to avenge wronged victims without addressing the cultural forces that encourage gendered violence (O'Toole 2013).

The second case study explores Philip DeFranco's response to the case. DeFranco is a popular YouTube vlogger who runs a satirical/comedy news vlog and who made one of the most-watched YouTube videos on the Steubenville[3] case. DeFranco's video challenges rape culture, but he is not a straightforward feminist activist: he has come under fire for sexist comments (Jaworski 2013) and describes himself as a 'sexist feminist' (DeFranco 2013c). We discuss Lostutter and DeFranco as examples of a burgeoning online frontier of bystander intervention, in which online counter-publics are used to challenge dominant rape narratives, particularly in spaces where feminist discourses have been resisted. This chapter offers a critical reading of how young men are portrayed within online discussions of rape by exploring how Lostutter and DeFranco sought to challenge rape culture within online hacktivist and gaming-centred spaces: spaces that are often characterised as misogynistic (Frank 2014; Massanari 2015).

Case study 1: Deric Lostutter

Deric Lostutter is a member of the online collective Anonymous that emerged out of 4chan[4] in 2003 (Coleman 2014). While the group was first associated with pranking and trolling, in 2008 Anonymous members (hereafter referred to as anons) shifted their focus and launched operations against the Church of Scientology (utilising techniques ranging from distributed denial-of-service [DDoS[5]] attacks to phone pranking), and since then anons have generated a multifarious range of activist subgroups (Coleman 2014). Prior to Steubenville, Lostutter created a viral 'operation'[6] targeted at Hunter Moore, the infamous host of the revenge porn site *Is Anyone Up?* He dubbed this offshoot of Anonymous, aimed at defending women, 'Knightsec'. This case study will provide a brief background to Lostutter's involvement in the Steubenville case, before discussing his self-portrayal, with a particular focus on interviews published in *Esquire* (Richardson 2013), *Gawker* (Chen 2013) and *Rolling Stone* (Kushner 2013). Through this case study, we explore male online vigilante rape activism and its complicated engagement with a feminist ethics, with a particular focus on the role of 'geek masculinity' and the logic of protectionism inherent to Lostutter's approach.

After the evening of 11 August 2012, Jane Doe learned about her assault through images that were shared on Twitter and Instagram. After she contacted the police on 17 August, Mays and Richmond were charged. There was initially limited media coverage of the case, with wider interest largely generated over social media, particularly through citizen activists Michelle McKee and Alexandria Goddard, who tweeted and blogged about the case. On 16 December 2012, the *New York Times* published a substantial exposé on the case (Macur and Schweber 2012). Across the country, Lostutter read the *New York Times* story and was perturbed, particularly by the alleged cover-up of the assault by members of the local community. He then tweeted about the incident from his Anonymous account @KYAnonymous, a tweet that was read by McKee. Recognising that the vigilante group may be interested in this case and that Lostutter had experience exposing similar cases in the past, McKee contacted Lostutter and asked him to help. After corresponding with McKee, in late December 2012, Lostutter made a YouTube video in the traditional Anonymous style – wearing a Guy Fawkes mask and using a computerised voice. He introduced the video, 'Greetings, citizens of the world. We are Anonymous. We are Knightsec', and in it he threatened to dox those complicit in the assault and the alleged cover-up unless they came forward and publicly apologised to the victim by New Year's Eve (Kushner 2013).[7] Lostutter named this operation OpRollRedRoll.[8]

Due to the loose structure of Anonymous, operations generated by one anon are taken up and contributed to by others. After Lostutter tweeted his threat, one anon hacked the Steubenville High School website. Lostutter's call for information about the assault was answered by others, and he received 'leads' about the case (of varying credibility), which he retweeted to his 30,000 followers (Chen 2013). While the aggregation and publication of images and videos from the night of the assault was a key part of the operation, it also crossed virtual borders. Knightsec helped organise a rally in Steubenville on 29 December at which 1,000 people gathered in support of the victim. On New Year's Eve, at Jane Doe's request, Lostutter did not execute his threat (he later claimed that the goal 'wasn't to hack . . . [but] to get the attention of higher authorities to get involved' and to 'scare' students into coming forward with information [Kushner 2013, p. 25]). Instead, he released footage of a 12-minute video (sent to him by another anon) of a former Steubenville football player, Michael Nodianos, joking about the victim being raped on the night of the attack.

A number of strategies are deployed by Lostutter, and we begin by considering his OpRollRedRoll video. In this video, Lostutter demands that perpetrators and bystanders come forward and apologise to the victim. In doing so, he engages in agenda-setting and framing the case according to his own politics of inclusion and exclusion: a politics that positions the victim as a member of a global community and bystanders as failing in their responsibility to protect and intervene. He later justified his interventions based on recognition of and sympathy for the victim and her experiences:

> I literally actually cried over what they did to this girl. . . . Here is this bright, lovely young girl whose life is forever changed because these people wanted to have a good night, it's bullshit.
>
> (Quoted in Chen 2013)

In Lostutter's response to the case, it is Jane Doe, not the perpetrators, who is discursively established as part of a community and whose potential and future are highlighted as valuable. Lostutter states that the operation was created out of his concerns about the lack of bystander intervention:

> The more I found out, the more angry I got . . . what really got me heated is her friends and everybody else's friends stood around and watched this shit happen, and nobody did a fucking thing.
>
> (Quoted in Kushner 2013)

In focusing on bystander inaction, Lostutter reconfigures communal responsibility as involving a social obligation to respond to sexual violence. This is a direct deviation from print media reporting on the case, which has reflected a different politics of community belonging. Media analysis conducted by Gutsche and Salkin (2015) found that reports emphasized that Jane Doe was not a Steubenville local. Furthermore, while she was de-identified throughout the coverage, such that she became unknown and unknowable, Mays and Richmond were constructed as 'known' within the community, with a particular emphasis on their status as promising athletes. The local football team was positioned as a financial and cultural touchstone in Steubenville (cf. Macur and Schweber 2012) and Richmond and Mays as fallen heroes (Moody-Ramirez, Lewis and Murray 2015, p. 10). Media reports also focused on the Steubenville 'community' (the town and its people) as victims of the Knightsec campaign (Gutsche and Salkin 2015). In an article in *Jezebel*, the rape allegations (as opposed to the rape itself) were presented as a community trauma (Baker 2013, para 8). The myth of community evoked here has political implications: the wound that had been exacted on the community is interpreted not as the violence of rape but the violence of rape accusations. In this way the victim is further othered, while the perpetrators and their fates are depicted as intertwined with those of the Steubenville community.

Rather than taking the common route of individualizing and demonizing the perpetrators, Lostutter asserts that bystanders share responsibility and positions those outside the local community and outside the law as having an obligation to act and react through online and offline activism. This reflects the shifts in advanced or liquid modernity replete with fluid identities and notions of community (Bauman 2000) which are not based solely on family, kinship and geographical ties but also on collective ideological solidarity. Such changing social patterns have dovetailed with the digital revolution. Technological advances have contributed towards dis-embedding the social structure and '"lifting out" . . . social relations from local contexts of interaction and their restrict[ions] across indefinite spans of time-space' (Giddens 2013, p. 21), which renders 'the spirit of commonality or communion that is found among networks of people' far more important than simply having a sense of place (Ward 1999, p. 98). While online spaces like social media do not invoke the nostalgic community ideals of *Gemeinschaft*'s strong sense of common identity, close personal relationships and an attachment to traditional and sentimental concerns (Tönnies 1887/2001), virtual technology 'enables the social networking of humans, brings people together and mediates feelings of virtual togetherness' (Fuchs, Hofkirchner, Schfranek, Rafll, Sandoval and Bichler 2010, p. 49). Whether online community can function as a replacement for more traditional forms of community is highly contested (Ferlander 2003). Nonetheless, social media platforms provide forms of virtual community involving sustained patterns of communication between large numbers of people which can generate counter-publics that facilitate social justice movements (Megarry 2014; Salter 2013). Online counter-publics allow unacknowledged victims spaces of recognition by addressing these issues within online peer communities unbound by spatial limits (Powell 2015) in ways that 'serve not only to create a forum for alternative viewpoints but rather to modify existing principles and values, and to challenge (and even disrupt or transform) the normative structures of the public sphere' (Salter 2013, p. 229). The discussion of perpetrators' bright futures and a community under siege reflects a powerful symbolic discourse that calls upon mythic ideals of 'community', obscuring the darker element of *Gemeinschaft*. By imagining an online community of solidarity and justice, in which Jane Doe is seen as valued and important, Lostutter's engagement with digital space provides unacknowledged victims and online activists with a space to contest and recalibrate the dominant mythology of community.

In harnessing online space's potential for developing counter-discourses, Lostutter specifically used the tools of digital technology. OpRollRedRoll focused on aggregating videos and images of the assault, tweets about the assault and information about those perceived to be involved then harnessing the 'scalability' of networked publics (boyd 2011, p. 46) to disseminate this information via Twitter. This was illustrated by the most galvanizing contribution of the campaign, when Lostutter released the 12-minute video of Michael Nodianos discussing the rape, which was shared across social media and drew particular public ire. Nodianos was not present at the sexual assault, but the video showed him laughing on the night about the events: 'They raped her more than the Duke Lacrosse team . . . she is so raped . . . she is dead . . . she's deader than Trayvon Martin . . . her puss is as dry as the sun right now . . . Trent and Ma'lik raped someone . . . this is hilarious' (Bacchus 2013). This dehumanising commentary demonstrates that there exists a deeply problematic rape culture extending well beyond the two perpetrators. Lostutter's release of this private video was one way to destabilise the foundations of these embedded attitudes towards women and gendered violence. Through exposing an account of the assault, Lostutter engages in a practice of utilising networked publics to 'critically witness' (Girling 2004; Vitis and Gilmour 2016) and expose acts of sexual violence that often go unseen and unheard. However, by focusing this exposition on bystander attitudes towards the assault, Lostutter also engages in a cultural critique of a different kind – directing the online audience's attention towards both the victim's ordeal and broader community attitudes around sexual violence. As Penny (2013, para 3) argues, the release of this video was 'rape culture's Abu Ghraib moment'. Here, sexual violence was brought *into* the global community through the critical witnessing of rape-supportive attitudes and the positioning of bystanders as complicit in a culture that fosters sexual violence. It marks a direct deviation from traditional news framings of sexual violence, which have a tendency to focus on the victim's decisions, appearance, assumed character and behaviours (Benedict 1992; Franiuk, Seefelt and Vandello 2008; Meyers 1996). Moreover, by encouraging people to come forward with information about the case, Lostutter utilises a core technique of Anonymous: encouraging collective action at a distance via social media and mobilising audience members to transform from passive to active (Fuchs 2015, p. 93). This approach relies on networked notions of community, and its radical potential is accentuated as a response to sexual violence, which has long been privatised and individualised.

While campaigning around the case by Knightsec was done online, real-world activism was also central to their efforts, and they helped organise rallies at Steubenville (Pitz 2013). These rallies provided an opportunity for a show of large-scale public support for women who had been raped and could share their stories and be supported by a mix of friends and strangers (Harkinson 2013) in a community spectacle of inclusion. The cultural performance of community is always contested and symbolic (Fuoss 1995), and in these public rallies, the acceptance of sexual violence within close-knit rural communities (Barclay, Donnermeyer and Jobes 2004) was further challenged. Lostutter's activism utilises discursive contestations around the meaning and practice of community and identity and the values and gendered subject positions underpinning them. In critiquing rape-supportive attitudes and bystander inaction and exposing these practices within online, global communities, Lostutter's response to Steubenville unsettles framings of rape within the public sphere by focusing on underlying cultural and structural issues and critiquing community complicity in the rape and its aftermath (Glenza 2015). While the focus of Lostutter's 'operation' centred on emphasizing communal responsibility, interwoven within this vigilante 'operation' was a performance of 'geek masculinity' and protectionism, which we explore in what follows.

Lostutter's online activism is both politically engaged and performative, and he invokes both the broader Anonymous persona and his own self-portrayal of working-class masculinity. *Rolling Stone*'s portrait of Lostutter paints him as the stereotypical 'geek'/'nerd' of the internet, which in the cultural imagination is 'first and foremost, a reclusive and socially awkward male immersed in the minutiae of computing' (Westman 2007, p. 26). While it has a fluid, even volatile, and loose organisational structure, Anonymous relies heavily on the symbolic and the visual in mobilizing resistance and participation (Lindgren and Stoehrel 2014). While there are undoubtedly female anons, Anonymous members tend to be discursively constructed as white and male (Coleman 2014), as are hackers more broadly (Tanczer 2016). MC (@Master_of_Ceremonies), a de facto spokesperson for the Anonymous effort in Steubenville, has observed of the Anonymous demographic:

> [They are] guys sitting around playing World of Warcraft and drinking beers [or] out on the streets where there's riots and kettles . . . it's just kind of more a guy mentality thing.
>
> (Harkinson 2013)

While Anonymous's operations are diffuse in their focus, there has been a recent shift towards operations targeted at perpetrators of gendered violence, argued by some to have begun in 2012 with Lostutter's operation against Hunter Moore. As a result, Lostutter has been credited with ushering in a new 'white knight' faction of Anonymous (Kushner 2013). The day after Moore's websites were taken down, Lostutter tweeted (as @KYAnonymous), '#Knightsec has been born . . . add it to your bio'. This shift in focus has been both lauded and perceived as unsettling. While online geek culture, including that which permeates Anonymous, might contest hegemonic masculinity, it has a fraught relationship with gender (Milner 2013). As Chun and Friedland (2015, p. 14) argue:

> given the pushback against feminists and rape victims on forums such as Reddit, we need to ask: why and how have rape victims become such a cause célèbre for groups such as Anonymous? And how can this create a link between two groups that are usually considered at odds: hackers and 'girls'?

As such, Lostutter's response, the rise of Knightsec and the use of hacking in the defence of women are located within a fraught cultural milieu that has been resistant to feminist practice. Such tensions are evident in examinations of Anonymous's interventions into gendered violence. In going after the perpetrators of sexual violence, Anonymous enforces a logic of protectionism that is explicitly gendered (Chun and Friedland

2015). Lostutter and Knightsec are repeatedly compared to superheroes. His involvement meant that he was 'caught up in one of the hottest battles of the modern world' (Richardson 2013). He was a 'self-ordained superhero with a new mask and a mission' (Kushner 2013). Richardson (2013) observed that '[o]nce you put on the Guy Fawkes mask and start talking through one of those voice-altering Darth Vader machines, you become something larger than yourself, an icon, a kind of superhero' and that the video Lostutter made 'felt like an outtake from *The Dark Knight Rises*'. Lostutter himself observes that '[t]he good guy stands up for the victim' (Chen 2013). Yet the cultural representation of the superhero remains highly gendered. As Stabile (2009, p. 87) observes in her analysis of the contemporary superhero:

> [Superhero narratives] are quintessential scenarios that indulge in fantasies about the heroes' unlimited ability to protect a silent and largely feminized humanity from that which threatens it. . . . While the masculine characteristics of the superhero/protector may vary – these men are as likely to present as mild-mannered or tormented in the contemporary moment as they are to appear as John Wayne–type cowboys – they are uniformly male.

This symbolic interpretation of this genre of online activism as a masculine battle between good and evil is further apparent in the articulation of such vigilantism as a form of violence. Lostutter refers to his activism as an effort to 'weaponize the media' (Chen 2013), while media reports have suggested that he 'declared war on Steubenville' (Richardson 2013). Lostutter observes:

> I don't really get down with violence toward women. . . . And the people I used to run with and hang out with, and how I was raised, if you did this at my house, or the house I was at, I was going to kick your fucking ass. Period. I was going to take you outside and beat the hell out of you. I can't do that over the internet, so I did the next best thing.
>
> (Chen 2013)

The language adopted is thus one that emphasises and glorifies masculine violence (including online violence) when it is used to defend women. Such protectionist evocations evident in Lostutter's sentiments can be criticised for envisaging men as 'knights in shining armour' coming to avenge wronged (female) victims. While this critique needs to be complicated by the emphasis on bystander and community detailed earlier, this hero-based desire to seek justice may account for some of the problematic outcomes of the case such as the exposure and dissemination of Jane Doe's identity by some anons (Coleman 2014) and baseless accusations by other anons that a Steubenville High School webmaster's emails contained instances of child pornography[9] (Kushner 2013). This high-handed approach and the lack of consultation with the victim before the threatened dox illustrate some of the tensions between hacktivism and feminist practice, particularly as regards the primacy placed on the victim and their needs (Renzetti 2013, p. 46).

This complex manifestation of activism is located within an online environment that simultaneously supports and contests operations that focus on VAW. When Lostutter began doing social-justice activism, internet trolls accused him of 'moral-fagging, their term for any behavior that aspires higher than prankish nihilism' (Richardson 2013). In using the term 'Knightsec' to designate Anonymous operations targeted at perpetrators of gendered violence, Lostutter is offering a joking retort and plays with the concept of the 'white knight'. The name itself is a reappropriation of the slur 'white knight' used by internet trolls to mock do-gooders online by asserting that white knights are weak men who only defend girls in order to have sex with them (*abstractsplash* 2009). By re-imbuing the concept with a more heroic character, Lostutter is simultaneously reinvigorating a gendered logic of protectionism and also resisting the disciplinary techniques within this broader online community that attempt to shame and ridicule those who seek to use technology to address VAW. While the language of superheroes and violence suggests a response that is symbolically coded as unequivocally masculine and reiterative of paternalistic vigilantism, the actual engagement signalled something more complex than the defence of virtuous women against evil men. The perpetrators were targeted, but so were the structural and cultural forces that support rape culture.

Case study 2: Philip DeFranco

Philip DeFranco is a YouTube vlogger who makes comedy news and current-affairs vlogs. He currently has over 4.8 million subscribers and 1.5 billion views.[10] His show is aimed primarily at young men: it frequently involves pictures of women in their underwear and has a focus on gaming and tech toys. DeFranco does not just embody 'geek masculinity', he is explicitly part of the gaming community, which, particularly post-#Gamergate,[11] is stereotyped as a horde of angry, misogynistic, young, white men. 'Geek' masculinity both draws upon and resists hypermasculinity (Massanari 2015) in valorising intelligence and rationality to the exclusion of emotionality while de-emphasising physical prowess (Kendall 2000; Massanari 2015). Such masculinities have been identified in online spaces imbued with nerd culture such as subreddits,[12] 4chan, 8chan (Salter 2017), video games (Jenson and De Castell 2010) and multiplayer online games (Braithwaite 2014). Within online gaming spaces, disdainful and dismissive attitudes towards women gamers are prominent (Salter and Blodgett 2012), vitriol has been directed at women who critique sexist games and gaming communities (Frank 2014), women have been constructed as 'objects of desire or unwelcome interlopers' (Massanari 2015: 5) and feminism has been framed as a 'threat to community stability and sociality' (Braithwaite 2014: 715). DeFranco's own videos that discuss gender issues frequently attract anti-feminist comments from his followers and diatribes against SJWs ('social justice warriors').

On 13 March 2013, DeFranco published a video called 'The Steubenville "Rape Crew" and their slutshaming supporters' (DeFranco 2013a) in which he questions neoliberal conceptions around gendered risk and responsibility and criticises those he sees as 'slutshaming'. First, DeFranco refutes the notion that drinking renders women partially responsible for their victimisation. He directly addresses the issue of intoxication and consent: 'She goes to a party and she gets really, really drunk. At first she's a little sloppy, then she's slurring, then she's non-responsive. We know this because people posted pictures of it on the Internet'. DeFranco articulates that the case is therefore categorically a rape: 'If you think no, that isn't a rape, then that's probably you trying to justify something. That's a rape'. This challenges double standards of responsibility when it comes to intoxication and sexual assault, particularly that intoxicated perpetrators tend to be held less responsible for their actions, while intoxicated victims are held to be more so (Finch and Munro 2007). Traditional media constructions of rape involving alcohol construct the issue as a problem of female drinking rather than one of male violence (Meyer 2010). Responses to Steubenville were no exception, with newspaper reports focusing on the victim's alcohol consumption. In response to 'slut-shaming and victim-blaming' commentators on the case, who argued that 'if a person put themselves in this situation and jumped into a pool of sharks, you don't blame the shark', DeFranco responded:

> Which in that situation, no, you would not blame the shark. But unless they [Mays and Richmond] said, 'Hey unnamed 16-year-old girl. Do you want to go to a rape party? It's the best rape party in town . . .' – that is not the same thing. Unless it's an understood fact that all men, 100% of men are rapists, you can't argue that.

Here, DeFranco refutes the discourse of risk that posits women's perpetual vulnerability to rape and men's natural sexual aggressiveness. While DeFranco's video was largely positively received, some of his viewers criticised his position, to which he responded in a Facebook post on the following day, 14 March 2013:

> I'm super disappointed in so many of you slutshaming a girl who was sexually assaulted when she was passed out drunk. . . . How dare a girl who thought that she was surrounded by friends who would protect her drink alcohol. Reevaluate your views on life.
>
> (DeFranco 2013b)

DeFranco acknowledges Jane Doe's drinking but does not render her responsible for her rape and refuses a cautionary tale of victim blaming while simultaneously deploying feminist concepts of slut shaming and victim blaming. He berates those who blame victims and constructs them as part of a broader societal problem. DeFranco, like Lostutter, is critical of the broader social and cultural context in which sexual assault occurs. DeFranco provides a link to the video of Michael Nodianos and describes how 'it just goes on and on with these stupid fucking terrible stomach-churning things'. He also criticises the broader Steubenville community: 'There's also all these tweets about the event, other people that weren't involved retweeting it. Really I'm amazed that there were only two people on trial because in Ohio it's illegal not to report a felony'. While critiquing the local community in which the assault occurred, DeFranco calls on his online community – a group largely hostile to feminism – to discuss the case. DeFranco also tries to elicit a sympathetic response to Jane Doe from his viewers:

> What you need to do if you're slut-shaming, if you're still on that side, is think – what if my future daughter, my current daughter, my sister, my mum when she was younger. What if this had happened to her? Would I be so dismissive?

While an entreaty to consider the case in terms of how we would feel about loved ones in order to encourage viewers not to dismiss Jane

Doe's experience, this framing clearly departs from a feminist ethics: it implies that a woman is valuable insofar as she is loved and valued by the men in her life. DeFranco's engagement here with a feminist ethics is complex: he draws on key ideas, such as slut shaming and victim blaming, in condemning the perpetrators and the culture that supports them while continuing to uphold the value of a masculine perspective. A key purpose of DeFranco's videos, however, is that they are meant to be funny, and the final section of this video involves a certain black humour:

> And I'm not trying to White Knight this thing, it's just don't be a shitty person. If these were my kids, I'd just tie my tubes and say: 'Well ok, I failed, I did a bad job and brought some evil into this world'.

As we have argued elsewhere (Vitis and Gilmour 2016), humour can be used as a form of activism to reveal the absurdities underpinning stereotypes (Bing 2004; Case and Lippard 2009; Merrill 1988; Weaver 2010; Zimbardo 2014). It makes common assumptions 'visible, and their stereotypical distortions laughable' (Zimbardo 2014: 61), and the entertainment quality of DeFranco's vlogs renders this clearly an occasion of the playful being political. Jokes about rape are problematic and have been the topic of intense debate (Holm 2016; Kramer 2011). DeFranco's humour minimizes sexual violence yet is also employed to challenge the phallocentric logic of rape culture. In this way, his comedy is aligned with a range of online activist projects that deploy humour in order to combat gendered harassment and rape culture (Rentschler 2014; Vitis and Gilmour 2016). DeFranco's strategic use of irony – 'the best rape party in town' – and his suggestion that he'd tie his tubes if his sons were perpetrators illustrate the use of comedy as a form of activism as well as speaking to the hyperbolic affectations of internet discourse.

Young people are much more likely to turn to comedy or satirical news sources than their predecessors (Pew Research Centre 2012). Hanson and Haridakis (2008) observe that many young people watch both traditional news and comedy/satire news online but that they watch the different formats for different reasons: those who watch more traditional news want information; those who watch news in comedy and satire formats want entertainment. DeFranco's efforts to engage his audience in discussions around gendered violence, however flawed, are thus aimed at an audience who, because of both demographic and intent, are watching the show to be entertained and may not be engaging with these issues through a feminist lens elsewhere. Moreover, the entertaining format in which it is delivered does not necessarily render the content less persuasive. Humour's potential for persuasion has met with ambiguous research results (Young 2008). Nonetheless, Xenos and Becker (2009) observe that extant research suggests that, in addition to entertaining viewers, political comedy appears to play a role in helping some individuals learn about the political world. Moreover, Young (2008) argues that when arguments are delivered in a humorous way, recipients are less likely to respond negatively, in a challenging and critical way. The means by which we can effectively engage men in responding to and preventing sexual violence is thus a contested and complex issue. Men frequently respond to discussions around gendered violence and sexual assault prevention in a defensive manner. There has been little discussion, however, of the potential for humour to connect with men and reduce male defensiveness (Rich 2010). Whether humorous message content can raise awareness and disseminate information in ways that minimise anger and resistance is an important area for further research.

Conclusion

In this chapter, we have argued that online social media produces a discursive space in which the 'cultural scaffolding of rape' (Gavey 2005) can be acknowledged and challenged. By troubling the various building blocks that support a rape culture, online media activists and agenda setters create a space in which counter-publics can emerge and undermine dominant rape scripts (Salter 2013). While technology has been used to further facilitate harmful behaviour (Henry and Powell 2014), we argue that there is also scope to explore the potential for online discussion and its impacts on attitudes to function as a late modern expression of bystander intervention, which draws upon a more fluid and online networked understanding of community. Online activism did not just bring the Steubenville case to the attention of national media, it also mobilised public sentiment around the issue. Commentators have suggested that the online response to the case was instrumental in shifting discourse around the issue. Clark-Flory (2015) quotes Katie Hanna, executive director of Ohio Alliance to End Sexual Violence, as stating,

> The teen response on social media initially during the Steubenville case was incredibly negative and very victim-blaming, but then we saw a shift throughout the trial and in the last two years. Many more people are intervening online to say, 'This isn't acceptable'.

Bystander intervention is complex, and intent and actions are not always the same. Nonetheless, for bystanders to intervene, individuals need to be able to recognise and be aware of sexual violence (Banyard and Moynihan 2011), be motivated to take action and feel responsible for doing so (Banyard 2008). As well as raising awareness of the Steubenville case, Lostutter's and DeFranco's activism involves defining sexual violence, motivating intervention and situating bystanders as responsible for preventing sexual violence and contesting rape-supporting narratives. DeFranco, in particular, focuses on the definition of rape. They both explicitly aim for a broader discussion around the issue, with Lostutter trying to recruit others into online and real-world activism and DeFranco encouraging his viewers to debate and discuss the issue. Moreover, both admonish those in Steubenville who did not help the victim, and both clearly feel a responsibility to do so. While there are obvious critiques of his style of vigilantism, Lostutter's approach was underpinned by the presumption of a broader social responsibility to seek justice for women who have been victims of sexual violence. His engagement thus reifies one of the key principles of bystander intervention: the need to shift the focus of and responsibility for sexual violence prevention away from victims and onto community actors (Powell 2011), particularly men who are positioned as productive players within this process.

Studies have consistently shown that young men overestimate other young men's sexism (Kilmartin, Semelsberger, Dye, Boggs and Kolar 2014). Fabiano, Perkins, Berkowitz, Linkenbach and Stark (2003) found that men underestimate the importance that other men place on consent and the willingness of other men to intervene against sexual violence. Furthermore, men's willingness to act as women's allies is strongly influenced by their perceptions of others' norms (Fabiano, Perkins, Berkowitz, Linkenbach and Stark et al. 2003) and hampered by their perceived potential to lose social standing and descend the hierarchy of masculinity (Carlson 2008). Online activism around the Steubenville rape case suggests that online space can also enable ongoing peer education around these issues. Whether or not this can be effective in changing young men's attitudes warrants further research.

As narratives about women's sexual suffering and vulnerability continue to dominate popular discourse around rape, online space offers the possibility of challenging such narratives through building alternative discourses (Loney-Howes 2015; Salter 2013) and wider communities of resistance (outside traditional activist circles) to deconstruct rape culture in ways that are dynamic and interactive. Lostutter's and DeFranco's self-driven yet community-situated hacktivism and blogging illustrate a departure from the dominant framing of young men in online space as solely complicit in rape culture in favour of a deployment of alternative frameworks of young masculinity. Their activism also illustrates the potential for online space to blur the lines between performance and activism and for practices of resistance to rape culture that do not follow a strictly feminist grammar, presentation or agenda. As Harp, Loke and Bachmann (2014) observe, the electronic media landscape has allowed for more diverse feminist perspectives about rape and sexual assault to emerge.

This chapter highlights the online practices of young men who are actively undermining cultural narratives about the inevitability of rape, the impact of rape on the community and sympathy for rape perpetrators. While they very much embody forms of stereotypical online masculinity, Lostutter and DeFranco both, in their own way, call for social change in response to sexual violence. These young men have a complicated relationship with a feminist ethics, valorising male intervention and male perspectives (respectively) while blurring the boundaries between activism and entertainment in a manner characteristic of social media engagement. As we have argued elsewhere, constructing young women as the hapless victims of online space is problematic (Vitis and Gilmour 2016). Similarly, constructing young men *solely* as contributors to or silent spectators of rape culture is also problematic. We concur with Salter (2017: 2729) that 'social media cannot replace struggles "on the ground" and its ties may be too weak to act as the sole catalyst of political mobilisation'. Yet if we accept a poststructuralist account that argues that rape culture not only excuses rape but is also a key component in creating the conditions in which rape occurs, then we need to explore alternative discussions of rape and the possibility for counter-publics to emerge in online space. In engaging with these issues online, Lostutter and DeFranco both certainly leave themselves open to feminist critique. However, we argue that interpreting resistance and bystander intervention as the sole province of those who fit a feminist ideal risks obscuring the more complex (and fraught) cultural shifts and forms of resistance that are apparent in young men's engagement with issues of sexual violence. However flawed their engagement with a feminist ethics, young men's own online engagement, and particularly that directed at their peers, must be seen as a crucial site of analysis.

Notes

1 Video blogger, someone who makes videos on YouTube, as a career and/or a hobby.
2 Expose their personal information, including names, social security numbers, addresses, names of relatives and phone numbers.
3 725,558 views at November 2016.
4 4chan is an image-based bulletin board where users post comments and share images anonymously.
5 Making online services unavailable by overwhelming them with traffic from multiple sources.
6 An operation that is circulated widely and rapidly on the internet, gaining support from other anons.
7 Video is available online at 'Steubenville Ohio Rape Case #Oprollredroll #Occupysteubenville' (JR_113 2013). Lostutter remained anonymous throughout the Steubenville case and trial. His identity was later publicly revealed.
8 'Roll Red Roll' is a chant used by the Steubenville football team.
9 The women in the hacked emails were all over 18. These allegations resulted in a Federal Bureau of Investigations investigation. Lostutter stated, 'I feel bad, if I could talk to the dude and say sorry, I'd tell him I'm sorry for putting his name out there and putting his business out there' (Chen 2013, para 54).
10 As of October 2016.
11 Gamergate refers to a diffuse social movement largely conducted on Twitter in which concerns about members of the gaming community harassed game developers Zoe Quinn and Brianna Wu and critic Anita Sarkeesian after Quinn's boyfriend accused her (via a blog) of having a relationship with a gaming journalist for positive reviews. The result of this blog was a targeted Twitter harassment campaign against these women,

which was framed by perpetrators as a manifestation of concern for 'ethics in gaming journalism' (Salter 2017).

12 Reddit entries are divided into categories of interest; each category is referred to as a subreddit.

References

abstractsplash (2009), 'Internet White Knight', *Urban Dictionary*.

Bacchus, M (2013), 'Michael Nodianos confession #oprollredroll #occupysteubenville', *YouTube*, available at www.youtube.com/watch?v=A0drRrNWpNE.

Baker, KJM (2013), '"A town destroyed for what two people did": dispatch from Steubenville', *Jezebel*, 16 September, available at http://jezebel.com/a-town-destroyed-for-what-two-people-did-dispatch-fr-1298509440 (accessed 5 July 2016).

Banyard, VL (2008), 'Measurement and correlates of prosocial bystander behavior: the case of interpersonal violence', *Violence and Victims*, vol. 23, no. 1, pp. 83–97.

Banyard, VL and Moynihan, MM (2011), 'Variation in bystander behavior related to sexual and intimate partner violence prevention: correlates in a sample of college students', *Psychology of Violence*, vol. 1, no. 2, pp. 287–301.

Barclay, E, Donnermeyer, JF and Jobes, PC (2004), 'The dark side of Gemeinschaft: criminality with rural communities', *Crime Prevention and Community Safety: An International Journal*, vol. 6, no. 3, pp. 7–22.

Bauman, Z (2000), *Liquid Modernity*, Polity Press, Cambridge.

Benedict, H (1992), *Virgin or Vamp: How the Press Covers Sex Crimes*, Oxford University Press, New York and London.

Bing, J (2004), 'Is feminist humour an oxymoron?', *Women and Language*, vol. 27, no. 1, pp. 22–33.

boyd, d (2011), 'Social network sites as networked publics: affordance, dynamics and implications', in Z Papacharissi (ed.), *Networked Self: Identity, Community, and Culture on Social Network Sites*, Taylor & Francis, New York, pp. 39–58.

Braithwaite, A (2014), '"Seriously, get out": feminist on the forums and the war(craft) on women', *New Media & Society*, vol. 16, no. 5, pp. 703–18.

Campbell, A (2005), 'Keeping the "lady" safe: the regulation of femininity through crime prevention literature', *Critical Criminology*, vol. 13, no. 2, pp. 119–40.

Carlson, M (2008), 'I'd rather go along and be considered a man: masculinity and bystander intervention', *The Journal of Men's Studies*, vol. 16, no. 1, pp. 3–17.

Case, CE and Lippard, CD (2009), 'Humorous assaults on patriarchal ideology', *Sociological Inquiry*, vol. 79, no. 2, pp. 240–55.

Chen, A (2013), '"Weaponize the media": an anonymous rapper's war on Steubenville', *Gawker*, 12 June, available at http://gawker.com/weaponize-the-media-an-anonymous-rappers-war-on-ste-512747826 (accessed 9 June 2016).

Chun, WHK and Friedland, S (2015), 'Habits of leaking: of sluts and network cards', *Differences: A Journal of Feminist Cultural Studies*, vol. 26, no. 2, pp. 1–28.

Clark-Flory, T (2015), 'Steubenville, two years later', *Refinery29*, 17 March, available at www.refinery29.com/steubenville-two-years-later (accessed 25 November 2016).

Coleman, G (2014), *Hacker, Hoaxer, Whistleblower, Spy: The Many Faces of Anonymous*, VersoBooks, London.

DeFranco, P (2013a), 'The Steubenville "rape crew" and their slutshaming supporters', *YouTube*, available at www.youtube.com/watch?v=CO-4xrTpffA.

DeFranco, P (2013b), 'I'm super disappointed in so many of you . . . ', *Facebook*, available at https://m.facebook.com/DeFrancoNation/posts/10151458109144407.

DeFranco, P (2013c), 'I think Im a sexist feminist . . .', *Twitter*, available at https://twitter.com/phillyd/status/348099706800443392.

Fabiano, PM, Perkins, HW, Berkowitz, A, Linkenbach, J and Stark, C (2003), 'Engaging men as social justice allies in ending violence against women: evidence for a social norms approach', *Journal of American College Health*, vol. 52, no. 3, pp. 105–12.

Ferlander, S (2003), *The Internet, Social Capital and Local Community*, Doctoral Thesis, University of Stirling.

Finch, E and Munro, VE (2007), 'The demon drink and the demonized woman: socio-sexual stereotypes and responsibility attribution in rape trials involving intoxicants', *Social and Legal Studies*, vol. 16, no. 4, pp. 591–614.

Franiuk, R, Seefelt, J and Vandello, J (2008), 'Prevalence of rape myths in headlines and their effects on attitudes toward rape', *Sex Roles*, vol. 58, no. 11, pp. 790–801.

Frank, J (2014), 'How to attack a woman who works in video gaming', *The Guardian*, 1 September, available at www.theguardian.com/technology/2014/sep/01/how-to-attack-a-woman-who-works-in-video-games (accessed 2 December 2014).

Fraser, N (1990), 'Rethinking the public sphere: a contribution to the critique of actually existing democracy', *Social Text*, vol. 25/26, pp. 56–80.

Fuchs, C (2015), 'Anonymous: hacktivism and contemporary politics', in D Trottier and C Fuchs (eds), *Social Media, Politics and the State*, Routledge, London, pp. 88–109.

Fuchs, C, Hofkirchner, W, Schfranek, M, Raffl, C, Sandoval, M, Bichler, R. (2010), 'Theoretical foundations of the web: cognition, communication, and co-operation – towards an understanding of Web 1.0, 2.0, 3.0', *Future Internet*, vol. 2, pp. 41–59.

Fuoss, KW (1995), '"Community" contested, imagined and performed: cultural performance, contestation, and community in an organized-labor social drama', *Text and Performance Quarterly*, vol. 15, pp. 79–98.

Gavey, N (2005), *Just Sex? The Cultural Scaffolding of Rape*, Routledge, London and New York.

Giddens, A (2013), *The Consequences of Modernity*, Wiley, Hoboken.

Girling, E (2004), 'Looking death in the face: the Benetton death penalty campaign', *Punishment & Society*, vol. 6, no. 3, pp. 271–87.

Glenza, J (2015), 'Steubenville rape cover-up charges dropped against school superintendent', *The Guardian*, 12 January, available at www.theguardian.com/us-news/2015/jan/12/steubenville-cover-up-charges-rape-dropped-superintendent (accessed 8 July 2016).

Gutsche, RE and Salkin, E (2015), 'Who lost what? An analysis of myth, loss, and proximity in news coverage of the Steubenville rape', *Journalism*, vol. 17, no. 1, pp. 456–73.

Hanson, G and Haridakis, P (2008), 'YouTube users watching and sharing the news: a uses and gratifications approach', *The Journal of Electronic Publishing*, vol. 11, no. 3.

Harkinson, J (2013), 'Exclusive: meet the woman who kicked off anonymous' anti-rape operations', *Mother Jones*, 13 May, available at www.motherjones.com/politics/2013/05/anonymous-rape-steubenville-rehtaeh-parsons-oprollredroll-opjustice4rehtaeh (accessed 25 November 2016).

Harp, D, Loke, J and Bachmann, I (2014), 'Spaces for feminist re(articulations): the blogosphere and the sexual attack on journalist Lara Logan', *Feminist Media Studies*, vol. 14, no. 1, pp. 5–21.

Henry, N and Powell, A (2014), 'The dark side of the virtual world: towards a digital sexual ethics', in N Henry and A Powell (eds), *Preventing Sexual Violence: Interdisciplinary Approaches to Overcoming a Rape Culture*, Palgrave MacMillan, Basingstoke and New York, pp. 84–104.

Holm, N (2016), 'Humour as edge-work: aesthetics, joke-work and tendentiousness in tosh.0 (or Daniel Tosh and the mystery of the missing joke-work)', *Comedy Studies*, vol. 7, no. 1, pp. 108–21.

Jane, E (2012), '"Your a ugly, whorish, slut": understanding e-bile', *Feminist Media Studies*, vol. 14, no. 4, pp. 531–46.

Jaworski, M (2013), 'Is Philip DeFranco a sexist or just misunderstood?', *The Daily Dot*, 6 August, available at www.dailydot.com/irl/philip-de-franco-sexist-women-laci-green/ (accessed 25 November 2016).

Jenson, J and De Castell, S (2010), 'Gender, simulation and gaming: research review and redirections', *Simulation and Gaming*, vol. 41, no. 1, pp. 51–71.

JR_113 (2013), 'Steubenville Ohio rape case #OpRollRedRoll #OccupySteubenville', *YouTube*, available at www.youtube.com/watch?v=1wfuy-vnpWY.

Kendall, L (2000), '"Oh no! I'm a nerd!": hegemonic masculinity on an online forum', *Gender & Society*, vol. 14, no. 2, pp. 256–74.

Kilmartin, C, Semelsberger, R, Dye, S, Boggs, E and Kolar, D (2014), 'A behaviour intervention to reduce sexism in college men', *Gender Issues*, vol. 32, no. 3, pp. 97–110.

Kramer, E (2011), 'The playful is political: the metapragmatics of internet rape-joke arguments', *Language in Society*, vol. 40, no. 2, pp. 137–68.

Kushner, D (2013), 'Anonymous vs. Steubenville', *Rolling Stone*, 27 November, available at www.rollingstone.com/culture/news/anonymous-vs-steubenville-20131127 (accessed 1 January 2016).

Lindgren, S and Stoehrel, RF (2014), 'For the lulz: anonymous, aesthetics and affect', *Journal for a Global Sustainable Information Society*, vol. 12, no. 1, pp. 238–64.

Loney-Howes, R (2015), 'Beyond the spectacle of suffering: representations of rape in online anti-rape activism', *Outskirts: Feminisms Along the Edge*, vol. 33, pp. 1–17.

Macur, J and Schweber, N (2012), 'Rape case unfolds on web and splits city', *The New York Times*, 16 December, available at www.nytimes.com/2012/12/17/sports/high-school-football-rape-case-unfolds-online-and-divides-steubenville-ohio.html?_r=0 (accessed 29 June 2016).

Marcus, S (1992), 'Fighting bodies, fighting words: a theory and politics of rape prevention', in J Butler and JW Scott (eds), *Feminists Theorize the Political*, Routledge, New York, pp. 385–403.

Massanari, A (2015), '#Gamergate and the Fappening: how Reddit's algorithm, governance, and culture support toxic technocultures', *New Media & Society*, pp. 1–18.

Megarry, J (2014), 'Online incivility or sexual harassment? Conceptualising women's experiences in the digital age', *Women's Studies International Forum*, vol. 47, pp. 46–55.

Merrill, L (1988), 'Feminist humor: rebellious and self-affirming', *Women's Studies*, vol. 15, no. 1–3, pp. 271–80.

Meyer, A (2010), '"Too drunk to say no": binge drinking, rape and the Daily Mail', *Feminist Media Studies*, vol. 10, no. 1, pp. 19–34.

Meyers, M (1996), *News Coverage of Violence against Women: Engendering Blame*, Sage, Thousand Oaks, CA.

Milner, RM (2013), 'Internet memes, identity antagonism, and the logic of lulz', *The Fibreculture Journal*, no. 22.

Moody-Ramirez, M, Lewis, T and Murray, B (2015), 'The 2013 Steubenville rape case: an examination of framing in newspapers and user-generated content, *Southwestern Mass Communication Journal*, vol. 30, no. 2, pp. 1–22.

Oppel, R (2013), 'Ohio teenagers guilty in rape that social media brought to light', *The New York Times*, 17 March, available at www.nytimes.com/2013/03/18/us/teenagers-found-guilty-in-rape-in-steubenville-ohio.html?pagewanted=all&_r=0 (accessed 25 November 2016).

O'Toole, E (2013), 'The anonymous problem with feminism', *The Guardian*, 7 December, available at www.theguardian.com/commentisfree/2013/dec/07/anonymous-problem-with-feminism (accessed 25 November 2016).

Pennington, R and Birthisel, J (2016), 'When new media make news: framing technology and sexual assault in the Steubenville rape case', *New Media & Society*, vol. 18, no. 11, pp. 2435–51.

Penny, L (2013), 'Laurie Penny on Steubenville: this is rape culture's Abu Ghraib moment', *New Statesman*, 19 March, available at www.newstatesman.com/laurie-penny/2013/03/steubenville-rape-cultures-abu-ghraib-moment (accessed 25 November 2016).

Pew Research Centre (2012), 'Section four: demographics and political views of news audiences', *Trends in News Consumption: 1991–2012: In Changing News Landscape, Even Television is Vulnerable*, Washington.

Pitz, M (2013), 'Crowd demands justice for Steubenville victim', *The Blade*, 6 January, available at www.toledoblade.com/Courts/2013/01/06/Crowd-in-the-hundreds-demands-justice-for-rape-victim-in-Steubenville.html (accessed 26 October 2016).

Powell, A (2011), *Review of Bystander Approaches in Support of Preventing Violence against Women*, Victorian Health Promotion Foundation, Melbourne.

Powell, A (2015), 'Seeking informal justice online: vigilantism: activism and resisting rape culture in cyberspace', in N Henry, A Flynn and A Powell (eds), *Rape Justice: Beyond the Criminal Law*, Palgrave Macmillan, Basingstoke and New York, pp. 218–37.

Rentschler, C (2014), 'Rape culture and the feminist politics of social media', *Girlhood Studies*, vol. 7, no. 1, pp. 65–82.

Renzetti, CM (2013), *Feminist Criminology*, Routledge, Oxon.

Rich, MD (2010), 'The interACT model: considering rape prevention from a performance activism and social justice perspective', *Feminism & Psychology*, vol. 20, no. 4, vpp. 511–28.

Richardson, J (2013), 'I am anonymous', *Esquire*, vol. 160, no. 4 (1 November), available at www.esquire.com/news-politics/a25210/i-am-anonymous-1113/ (accessed 25 November 2016).

Salter, A and Blodgett, B (2012), 'Hypermasculinity & dickwolves: the contentious role of women in the new gaming public', *Journal of Broadcasting & Electronic Media*, vol. 56, no. 3, pp. 401–16.

Salter, M (2013), 'Justice and revenge in online counter-publics: emerging responses to sexual violence in the age of social media', *Crime, Media, Culture*, vol. 9, no. 3, pp. 225–42.

Salter, M (2017), *Crime, Justice and Social Media*, Routledge, London.

Stabile, CA (2009), '"Sweetheart, this ain't gender studies": sexism and superheroes', *Communication and Critical/Cultural Studies*, vol. 6, no. 1, pp. 86–92.

Tanczer, LM (2016), 'Hacktivism and the male-only stereotype', *New Media & Society*, vol. 18, no. 8, pp. 1599–1615.

Tönnies, F (1887/2001), *Community and Civil Society*, Cambridge University Press, Cambridge.

Vitis, L and Gilmour, F (2016), 'Dick pics on blast: a woman's resistance to online sexual harassment using humour, art and Instagram', *Crime, Media, Culture*, DOI: 10.1177/1741659016652445.

Ward, KJ (1999), 'Cyber-ethnography and the emergence of the virtually new community', *Journal of Information Technology*, vol. 14, no. 1, pp. 95–105.

Weaver, S (2010), 'The "other" laughs back: humour and resistance in anti-racist comedy', *Sociology*, vol. 44, no. 1, pp. 31–48.

Westman, KE (2007), 'Beauty and the geek: changing gender stereotypes on the Gilmore girls', in SA Inness (ed.), *Geek Chic: Smart Women in Popular Culture*, Palgrave MacMillan, New York, pp. 11–30.

Xenos, MX and Becker, AB (2009), 'Moments of Zen: effects of *The Daily Show* on information seeking and political learning', *Political Communication*, vol. 26, no. 3, pp. 317–32.

Young, DG (2008), 'The privileged role of the late-night joke: exploring humor's role in disrupting argument scrutiny', *Media Psychology*, vol. 11, no. 1, pp. 119–42.

Zimbardo, Z (2014), 'Cultural politics of humor in (de)normalizing Islamophobic stereotypes', *Islamophobia Studies Journal*, vol. 2, no. 1, pp. 59–81.

6 The role of information and communication technologies in facilitating and resisting gendered forms of political violence[1]

Gabrielle Bardall

Introduction

Substantial gains have been made worldwide in enhancing women's participation in public life in the past two decades. In 1995, women comprised at least 30 per cent of parliamentarians in only five countries (2 per cent of the total), while today 46 nation-states have reached that threshold (IPU 2016).[2] Constitutional revisions and electoral reform globally have enfranchised women and facilitated their political participation by entrenching their rights, offering incentives and/or imposing sanctions on political parties and other public bodies to protect against gender-specific threats. Nonetheless, women generally remain acutely under-represented in parliaments (only 22.7 per cent of parliamentarians worldwide are women [IPU 2016]) and face deep-rooted obstacles to participation as voters and in other civil and public roles. Barriers range from inadequate or nonexistent legal protections to traditional cultural stereotypes and gender roles, lack of access to resources and civic education and generally lower levels of self-confidence among women in relation to pursuing public office and actively participating in civic life in other roles. Of all these barriers, political violence is perhaps the most insidious and affecting.

Focusing on the nexus of gender, violence and technology, this chapter chronicles the parallel rise in new information and communication technologies (ICTs) and women's increased political participation in developed and developing states. I argue that socio-psychological violence (such as acts of intimidation and moralistic shaming) is the most widespread and damaging form of violence against women (VAW) in politics and that this specific type of violence intensifies in online and other ICT spaces due to their unique ability to rapidly amplify the reach, impact and harmfulness of socio-psychological attacks. The chapter challenges and builds on the growing literature regarding political violence and women in both academic (Bardall 2013, 2016; Krook 2015; Krook and Restrepo Sanín 2016) and practitioner literature (Bardall 2011; iKnowPolitics 2007; Inter-Parliamentary Union 2016; National Democratic Institute [NDI] 2016; South Asia Partnership International 2006; USAID 2013). The discussion focuses in particular upon the ways in which ICTs are being utilised to inhibit women's political participation via gendered violence in a range of ways and in different jurisdictions. For women entering political life or holding public positions, new ICTs are frequently used as tools of gender-specific electoral and political violence. There is evidence of ICTs being used to perpetrate a broad range of violent acts against women during elections and in public life, especially acts that inflict fear and psychological harm. The specific characteristics of ICTs are particularly adapted to misuse in this manner. Despite these significant challenges, ICTs also offer groundbreaking solutions to respond to the gendered dimensions of electoral and general political violence. Notably, ICTs combat these forms of violence through monitoring and documentation, educational and awareness-raising platforms and empowerment and advocacy initiatives. The intersection of gender, violence and technology creates new ways to suppress women's public voices just as it opens new opportunities for their empowerment and expression.

Political violence and gender

Political violence is defined as 'violence aimed at achieving or resisting regime change in established power hierarchies and orders; asserting or resisting supremacy of one form of national identity over another or others; seizing and controlling economic, political or other resources in the form of mineral, key routes; or resistance to any of these forms of violence' (Breen-Smyth 2016, p. 569). Political violence violates an individual or group's civil and political rights to participate in and contribute to the process of governing their society through the use or threat of coercive force. It is defined as political rather than interpersonal or other violence by the nature of the objectives, the targets of attacks, the organisational structure of the groups responsible and the repertoire of actions (see European Consortium for Political Research website; Waters et al. 2005). Within the dominant literature on political violence, gender has been the subject of limited analysis. Before outlining how the key concept of gendered political violence is used in this chapter to explore technology's role in spreading or resisting violence, I outline the three key concepts that are often utilised in relation to this broad area of research and offer an overview of their strengths, limitations and points of intersection or overlap.

First, traditional political violence is *gender-differentiated* (GDPV) in its manifestations. Both women and men are victims (and perpetrators) of these types of violence; however, the types of acts committed, their frequencies and the actors involved in them are patterned according to the sex of the victim (Bardall 2011, 2016). Likewise, the micro-geography of political violence (that is, where violence occurs) is frequently correlated with the gender of the victim (Bardall 2011, 2016) and, like other forms of violence (such as intimate partner violence), it can take both physical and non-physical forms (see Bardall 2011, 2016; Krook 2015; Krook and Restrepo Sanín 2016).

In contrast, the second concept, *gender-based* political violence (GBPV),[3] exists where harm is committed that violates an individual or group's political rights on the basis of their gender-identity. GBPV comprises all violent infringements of civil and political rights that are directed against an individual or group because of their gender identity or that disproportionately affect a given gender.[4]

The third concept is a subset of GBPV: where GBPV specifically targets cisgender women in order to enforce patriarchal control of democratic institutions, it may be described as violence against women in politics (VAWP). This chapter focuses on female victims of violence in the case of both gender-specific and non–gender-specific acts of political violence. However, the use of the term 'violence against women in politics' in this chapter differs substantially from its use elsewhere in the literature (where it is commonly abbreviated as VAWIP). The concept of VAWIP in the literature is conflicted and both overly broad and excessively narrow. It has been used in conflicting manners, with confusion over the objects of violence,[5] the relationship of the concept to other related concepts[6] and its motives.[7] In recent writing, VAWIP has emerged as an excessively narrow notion, focusing exclusively on candidates and elected officials[8] and women as victims.[9] Narrow interpretations also inaccurately restrict perpetrators' objectives[10] and the context[11] of the violence. Although VAWP, according to my definition, is theoretically limited to gender-motivated violence, in practice, the concept is often employed indiscriminately in referencing any act of political violence in which a cisgender woman is the victim. The over-extension and misuse of the concept of VAWIP diminish its usefulness, while its focus on female candidates and elected officials diverts our attention from the other, key victims at the intersection of politics, violence and gender – namely men (Bardall 2011, 2016). In light of this, this chapter grounds its theoretical framework in the core distinction between the types of violence (GDPV and GBPV) based on the motivation or objective of the act. VAWP (as opposed to VAWIP as used elsewhere in the literature) is used here to describe those acts of gender-motivated political violence perpetrated against women *as women* to maintain patriarchal control of the political or electoral sphere. GDPV and GBPV (and their subtypes) share potential categories of victims (such as voters, candidates, elected officials, political journalists, election workers and civil servants), perpetrators (such as political opponents or co-party members, family and intimate partners, community actors, rebel groups and hackers), as well as the locations (private, public, domestic and virtual) and types (physical and non-physical) described earlier.

Each of these three concepts (GDPV, GBPV and VAWP) is critical to understanding political violence perpetrated through the medium of ICTs. Violence that takes place in virtual locations is, by definition, non-physical and is composed of various forms of threats, intimidation, character attacks, hacking, theft of personal information and personification. As discussed in what follows, many of these attacks are gender specific, attacking female party supporters, voters, candidates, elected officials and others by employing sexualised or gender-specific messaging with the objective of diminishing or coercing women's political voice (acts of VAWP). This chapter finds that VAWP that occurs in the form of cyberattacks uses sexually demeaning and threatening messaging to violate the civil and political rights of its victims. In instances of GDPV, women and men are indiscriminately victims of virtual acts of violence within the context of broader political pursuits. A gender differential may exist according to country-specific contextual factors regarding gender gaps in knowledge and access to and usage of ICTs.

For example, in countries where the digital fluency gap is large, women are more vulnerable than men to digital attacks. While the scope of this research does not extend to quantify this differential, it recognises its existence and describes some of the channels it takes, setting a course for further research.

This chapter is specifically interested in VAWP and the gender differential in non–gender-specific acts of political violence and how recognising the intersection of this form of gendered violence with ICTs offers new ways to understand political violence, its gendered nature and the extent to which ways to harness technology can combat this violence. It begins first by outlining the empirical research that informs the discussion and then examines the role of ICTs in both facilitating and resisting gendered forms of political violence.

Methodology

My argument for the existence and typology of gender differentiation and gender specificity in political violence is derived from a cross-national empirical analysis (Bardall 2011, 2016). Drawing on data covering more than 2,000 individual incidents of electoral violence collected by community monitors in six countries[12] between 2006 and 2010, I found that election violence has distinct gender dimensions in terms of the types of violence (physical or non-physical), the perpetrators involved (institutional, non-state or societal) and the victims targeted (politically affiliated, institutional, professional or private non-state or non-political actors; Bardall 2011, 2016). Most importantly for this chapter, this study indicated the existence of variation in the location of political violence, including violence that takes place in public, private, domestic and virtual locations. Although these 2,000+ data points suggested the existence of gendered political violence in virtual spaces, the incident-based data was insufficient to explore the breadth and variety of this intersection between technology, gender and violence. As such, this chapter takes the typology identified through the empirical analysis and applies it to technology-based violence observed in empirical examples drawn from the author's fieldwork with various international and non-governmental organisations in some three dozen countries in Sub-Saharan Africa, the Middle East/North Africa, Central Asia, the Caribbean and Southeast Asia, as well as interviews with social media and gender entrepreneurs from 2011 to 2016 and supplemental desk research. Drawing upon this typology and empirical examples, I consider the range of victims, perpetrators and types of violence that are engaged when technology is employed as a means of perpetrating VAW in political contexts. In the final section of the chapter, I also look at how ICTs are used to resist and prevent political violence.

ICTs and social media as implements of gendered violence

As women's participation in politics continues to grow internationally despite ongoing legal barriers and cultural resistance, women's vulnerability to political violence increases (IPU 2016), including in online spaces and via ICTs. The

United Nations (UN) estimates that 95 per cent of aggressive behaviour, harassment, abusive language and denigrating images in online spaces are aimed at women, most often by a current or former partner (United Nations General Assembly [UNGA] 2006). Karen Banks noted more than 10 years ago that '[t]he internet is not creating new forms of crimes against women . . . but it is creating new ways and means for crimes to be perpetrated' (Banks 2001, pp. 147–73, see also Kee 2005). The focus of this discussion is on examining this claim while considering the implications of these new ways and means of enacting violence in response to or to prevent women's political participation. In particular, the internet and other social media and ICTs have proven to be uniquely dangerous instruments in perpetrating political violence against women because of the relative importance of psychological violence in women's political experience. In my initial cross-national study of election violence in six countries, the proportion of non–ICT-based intimidation and psychological acts of violence experienced by women (threats, harassment, aggressively abusive or denigrating language) was nearly three times that experienced by men (ratio of 28:10 per cent), while men experienced more than three times the levels of physical violence as did women (39:11 per cent) (Bardall 2011, 2016). In this chapter, we delve further into this finding on the disproportionate impact of psychological violence on GDPV and VAWP to examine violence perpetrated through ICTs (which is, by definition, non-physical and psychological in nature and thus uniquely susceptible to threatening women's participation based on these findings).

ICTs may be used directly as a tool of intimidation by threatening or inciting physical violence against women candidates, voters or representatives. Such cyber harassment or intimidation includes sending abusive, threatening or obscene emails, explicit threats of physical and/or sexual violence and encouraging strangers to physically harm the victim, which in some cases results in actual physical assault (Citron 2014; Moolman 2013). Citron notes that, unlike cyberattacks against men, cyber harassment of women is particularly 'sexually demeaning . . . sexually threatening . . . reduces the victims to basically their sexual organs, and sends the message that all [the victims are] there for is to be sexually abused, used and thrown away, that they offer nothing' (Citron 2014).

Acts of VAWP may involve spreading reputation-harming lies, electronic sabotage in the form of extensive spam and damaging viruses, impersonating the victim online and sending abusive emails or fraudulent spams, blog posts, tweets and other online communications in the victim's name or subscribing victims to unwanted email lists, resulting in hundreds of unwanted messages daily (Citron 2014; Ellison and Akdeniz 1998; Kee 2005; Moolman 2013). Such attacks can be perpetrated by both strangers and individuals known to the victim, as well as by proxy stalkers and 'cyber-mobs' (Citron 2014, p. 5). They may either be perpetrated as a form of VAWP or have a distinct impact on female victims due to the sexualised nature of their content despite being perpetrated for purely political objectives (as in the case of GDPV).

Cyber harassment can result in serious harm to the victim, as in one of the earliest cases recorded – that of Cynthia Armistead, an American woman who received thousands of offensive messages and threats in 1996 after her stalker published false online advertisements offering her services as a prostitute and providing her home address and personal telephone number (Bocij 2004). More innovative and sophisticated forms of ICT-based attacks on women have been documented and include the use of 'spy software' (spyware enables users to have access to all keystrokes made on the computer, including all email correspondence, web surfing and internet communication); wireless technology to monitor private conversations; hacking; saved 'cookies' and browser histories; email tampering and interception; and visual surveillance and geographic tracking via global positioning system (GPS) software (Southworth, Finn, Dawson, Fraser and Tucker 2007). These ICT-based attacks have an overwhelming impact on women's private and professional lives. Indeed, some surveys estimate that over 80 per cent of victims in cyberstalking incidents are women.[13] Likewise, women are often more vulnerable to these attacks: in the developing world, women have less access to the internet and ICTs than do men.[14] And where women do have access, they tend to have less experience with and lower levels of education in relation to ICTs, making them less able to prevent and respond to cyberattacks, including hacks and leaks.

The use of these and other forms of ICT-based violence has been documented in cases of VAWP. During the post-election violence that occurred in Kenya in 2008–9, tribal-based political partisans sent SMS messages to women in opposing tribal-based political groups, threatening bodily harm, rape and even death (Muthoni Wanyeki 2010). Graphically violent tweets were used to make rape and murder threats against British Member of Parliament Stella Creasy and other prominent British women at a rate of up to 50 threats per hour over the course of 12 hours following them declaring their support for a particular feminist issue (Döing 2013). During her 2008 electoral bid, Hillary Clinton was the object of a video game in which players slap her each time she speaks.

Beyond these exploitative uses of the medium, certain qualities of social media make them peculiarly suited to inflicting psychological violence on women in civic life. The disproportionate impact of such violence on women stems in large part from women's unique vulnerability to attacks based on gender-specific cultural expectations of propriety and morality. By breaking into a traditionally masculine field that is frequently associated with rough behaviour and corruption, women are exposed to sexualised and/or morally degrading criticism that may carry additional weight due to social expectations of women's roles and moral character. Derogatory accusations of being a prostitute, a lesbian or otherwise sinful and/or sexually deviant are frequently levelled against women running for or holding office in many countries (Bardall 2011, 2016; NDI 2015). These 'morality-based' attacks often carry much greater social costs for women than for men because of the existence of double standards around what constitutes 'moral behaviour' for male and female politicians. The specific nature of social media plays to these imbalances and exacerbates attacks on women in public life in several ways.

First, the nature of messaging in social media facilitates ridicule, shaming and other psychological forms of violence against women in elections and in politics (both non–gender-specific and GBPV). The most effective social media messages are generally short (in the case of Twitter, limited to 140 characters), written in simple language and often humorous. A study of media coverage of Hillary Clinton's and Sarah Palin's 2008 political campaigns in the US found that the crudest attacks were online, including more than 500 YouTube videos under the search 'Hillary' and 'bitch', and multiple Facebook groups with obscene or sexist names, including the most popular (with 41,025 followers in March 2008) 'Hillary Clinton: stop running for president and make me a sandwich' (Falcone 2008; Jamieson and Dunn 2008). Crude and sensational messages circulate widely without the legal or professional ethical requirements imposed on traditional media to ensure accuracy, check sources and rectify errors. With extremely low barriers to entry, social media users may engage in character assassination at virtually no cost and with little personal consequence.

Social media also facilitates attacks on women's ethics and morality through the ubiquitous presence of images. The use of stereotypical or demeaning images and photos to sexualise, emotionalise and trivialise women poses a strong disincentive for women considering running for office and may even pose a direct threat to their personal safety (Blackman-Woods 2013). Women MPs in many countries report feeling compelled to be hyper-conscious about their appearance and physical posture in public due to the ubiquity of cell phone cameras.[15] Candid shots taken at unguarded moments – such as awkward or unattractive poses; eating, coughing or yawning; minor wardrobe malfunctions; and accidental exposure of skin/cleavage – and immediately posted and disseminated online to mock and ridicule can have a degrading and intimidating impact on women candidates and MPs. Exacerbating the issue, the ease of using programmes such as Photoshop allows perpetrators to modify snapshots or create entirely new images designed to denigrate, compromise or shame their victim. For example, numerous such images of Hillary Clinton were distributed on Twitter and Facebook during the 2016 election campaign.[16] In the case of female politicians, this is commonly manifested through sexually suggestive or demeaning images. On YouTube, videography is often matched with music and can be used to promote violence towards women in politics. Exemplary here are a YouTube music video that was posted during the 2008 US campaign that flashed photos of Hillary Clinton during debates as the lyric 'I'll beat that bitch with a hit' was repeated (Jamieson and Dunn 2008) and the widespread pornography spin-offs regarding former vice presidential candidate Sarah Palin. This form of violence, known as malicious distribution, uses technology as a tool to manipulate and distribute defamatory and/or illegal material related to the victim (Baker, Campbell and Barreto 2013). Women who accede to public positions are often in the line of fire for this type of attack.

The speed with which information travels through social media networks and the scope of its diffusion magnify the impact of acts of GDPV and VAWP. Retweets, shares and 'Likes' spread degrading, humiliating or threatening attacks on women in politics, civil activists and political supporters with almost uncontrollable rapidity. The scope of online stalking and harassment is likewise amplified (Arya 2013). In a recent case, in a single night, British Labour MP Jess Phillips reported receiving more than 600 rape threats and was subjected to some 5,000 Twitter notifications of people discussing whether they would sexually assault her (Hughes 2016). While the speed of internet attacks affects both men and women, the specifically sexualised nature of attacks against women amplifies and distorts their gendered messages.

Available redress for this type of attack, including community censure, website moderation and legal intervention, frequently takes effect only after the damage to the victim has been done. Self-policing functions of websites such as Facebook and Twitter are often weak and/or vulnerable to gender bias (see Al Jazeera 2013; Davidson 2013). Interventions may interrupt or halt a behaviour but less frequently correct false accusations or degrading images. Indeed, given tight electoral deadlines, harm to a victim's public image may be difficult or impossible to correct before ballots are cast. The reach of any given message on social media is dependent on the voluntary diffusion of the message by social media users. Therefore, efforts to rectify degrading depictions cannot be broadcast consistently to all consumers of the original message. Finally, the speed and scope of social media attacks have a particularly chilling effect on political aspirants, especially women, who are entering politics for the first time. In my research, I found that women frequently cited the threat of widespread, rapid public attacks on their personal dignity as a factor deterring them from entering politics.

In contrast to many forms of ICT-based VAWP that target a woman's public image, ICTs may also cause harm through their ability to silence and bury women who otherwise seek to build a public presence for political aims (described as 'symbolic violence' by Krook and Restrepo-Sanin 2016). Some recent cases demonstrate such explicit attacks on women's access to and visibility via ICTs. Between 2010 and 2013, Indian villages in Bihar and Uttar Pradesh states and the Priyadarshini Indira Gandhi Government College for Women in Haryana banned single (or undergraduate) women from using cell phones (single or undergraduate men were not affected). In 2004, the Saudi Arabian Ministry of Education banned women from carrying camera phones. Polls in Nigeria have registered support for banning women's use of mobile phones (2013). Control over women's access to ICTs is also reported extensively as a tool of domestic violence and includes privacy invasion through SMS stalking, monitoring ICT use and withholding permission from female family members to use cell phones (Madanda, Ngolobe and Amuriat 2009). In these cases, gender-motivated control of the use of ICTs intersects with VAWP where the forcible restriction of access to ICTs impedes women's civic or political participation by limiting their access to information, violating their privacy or, for candidates and aspirants, limiting their ability to be politically visible.

Additionally, violence perpetrated through social media benefits from a significant degree of legal and moral impunity. Legally, there is often a fine line between actionable harassment and abuse that cannot be regulated (including online 'trolling') (Citron 2014). Non-criminal offenders may feel a certain moral impunity as a result of being distanced from the victim. In social media, the perpetrator may never meet his victim in person and never see the impact of his acts, thereby enabling him to dehumanise the victim. Public figures, in particular, may be seen as 'fair game' for harassing and abusive behaviour because of this effect. Likewise, attacks on public figures gather more popular feedback online than do attacks on private individuals, and offenders may develop a sense of moral approbation for their acts based on the number of 'Likes' and re-shares in the cybersphere. Should they fear censure, the perpetrator may choose to remain anonymous to either their immediate community or general society. A cyberstalker or other online perpetrator can conceal their identity to a degree impossible in any other context by using different internet service providers and/or by adopting different screen names. More sophisticated perpetrators can use anonymous remailers to virtually erase their association as the source of an email or other online communication (Munyua, Mureithi and Githaiga 2010). Since attacks on public figures tend to attract particular scrutiny by law enforcement and media, the ability to perpetrate acts of violence anonymously has strategic value and a particular cost value for perpetrators of political violence.

The sense of impunity surrounding social media–based harassment and aggression may also be amplified because these acts lack identifiable leadership. An 'incident' of VAWP on social media is different in nature from a traditional act of violence in which the perpetrator is clearly identifiable. Instead, an incident of online violence is a collective phenomenon and may involve dozens or even thousands of 'perpetrators'. Terrifying for the victim, this is particularly empowering for the authors of the violence. Perpetrators may gain confidence and feel social approbation when their messages are shared, re-tweeted or 'Liked' on the internet. Given this anonymity, social media users may feel diminished accountability when they promote hurtful messages through their networks.

A final reason why social media–based violence is problematic for women in politics is the difficulty of regulating and punishing attacks. The realm of social media is one of relative legal impunity for the perpetrators of electoral violence against women. Legal protections defining gender-based violence (GBV) and sexual crimes are lacking or entirely absent in many countries. This gap is compounded by the even greater gap in the realm of cybercrime in many countries (Madanda, Ngolobe and Amuriat 2009). Common protections against ICT-based violence against women may be limited to defending against stalking and harassment through telephone calls and electronic mail (Essof 2009). Only two countries, Mexico and Bolivia, have specific legislation addressing violence against women in elections.[17] Accessing justice for women is similarly challenging and, for women who do successfully bring their cases to court, favourable rulings and enforcement of criminal sentences or penalties often prove elusive. Electoral violence perpetrated through social media channels is thus virtually impossible to limit or prosecute.

Resisting VAWP through ICTs

ICTs are also tools of empowerment for women entering politics and seeking to combat all forms of VAWP, especially that based in social media. In 1995, at the

Fourth World Conference on Women, the Beijing Plan of Action called on all states, as well as media systems and associations, and non-government organisations (NGOs), to increase the participation and access of women to expression and decision-making in and through the media and new technologies of communication (UN 1995). Almost 20 years after Beijing, social media is being used to combat VAWP through three main areas: (1) monitoring and documenting VAWP, (2) educating and awareness-raising and (3) empowerment and advocacy initiatives.

Monitoring and documenting GBV are notoriously challenging due to the intimate and often humiliating nature of the violence and fear of retribution. In the case of VAWP, these factors are compounded by the lack of awareness of the link between political violence and GBV and the perceived need of women candidates to publicly 'save face' by hiding their experience of violence. Yet baseline data on the prevalence of VAWP is vital to raising the profile of the problem, aiding its victims and identifying appropriate solutions to mitigate and prevent it in the future. ICTs are making major contributions towards overcoming some of these challenges and establishing critical documentation of the problem. Traditional sources of documentation for election monitoring and observation missions (EOMs) have expanded and are now able to more effectively monitor social media traffic thanks to the introduction of low-cost or public-domain software services such as Hootsuite, TweetReach, Klout, Social Mention and many others. As the issue of VAWP becomes more widely recognised and mainstreamed in election observation, these tools will enable EOMs to document incidents of social media–based violence and analyse their trends.

ICTs also facilitate the collection of data on acts of VAWP perpetrated 'offline' (that is, acts of physical, psychological and sexual election violence occurring in physical locations). Open-source platforms such as the Carter Center's election monitoring platform (ELMO) and the Kenyan-based Ushahidi facilitate the monitoring of violence and the transmission, analysis and mapping of incidents to enable the identification of patterns, trends, frequencies and hotspots of violence. Crowdsourcing tools collect information and data on incidents from the public via SMS, Twitter, Facebook, YouTube, phone calls and email, which is transmitted to web platforms and visually mapped using publicly accessible maps such as Google Maps. Numerous initiatives exist to geographically map experiences of sexism, crime and violence such as HollaBack, Everyday Sexism and HarassMap. These can be effectively engaged to document VAWP at the community level, during rallies, protests, polling and other collective events.

Most critical of all, victims of violence are increasingly able to report acts of VAWP without fear of physical retribution or public shame, thanks to the anonymity of ICTs (APC 2013, Chaio 2011). As demonstrated in Ushahidi deployments in Egypt and Syria, where the Ushahidi platform has been used specifically to document GBV (via HarassMap), women feel empowered to speak out safely about their experience of violence when they are able to do so quickly (through their cell phones) and anonymously. Users have testified to the sense of empowerment provided by being able to securely yet publicly denounce assaults on their dignity

(Chaio 2011). Chaio relates various examples from the Tahrir Square protests that occurred in Cairo in 2011, where women (and at least one man) were able to send a text message from a private location (such as a bathroom) to report sexually abusive incidents during the protests, away from the view of family members who might otherwise have prevented or sanctioned them or perpetrators who may have threatened them further. The empowerment effect was noted in at least one case, where the victim identified herself voluntarily to HarassMap and stated that she would not have otherwise reported the incident through traditional means for fear of being shamed.

These empowering aspects of ICTs, combined with the growing recognition of the human rights dimension of these forms of violence (Ballington 2016), are enabling local and international actors to reliably document and denounce gender-differentiated and gender-based political violence to international human rights treaty bodies and other UN structures. The Commission on the Status of Women (2013) and the UN Working Group on Discrimination Against Women in Law and Public Life (2013) have begun recognising technology-related forms of VAW. As data becomes more available on the political dimensions of VAW, further action can be taken. One example of this is the Carter Center and the Centre for Civil and Political Rights' joint program in Burundi, which uses the ELMO platform to collect data on GBV/GBPV and to disaggregate general political violence and human rights data for submission to the Convention on the Elimination of all Forms of Discrimination Against Women (CEDAW) review process.

Rapid response to mitigate VAWP and early warning to prevent it are both enabled through individual ICTs and powerful mash-ups such as Ushahidi. Social media monitoring software for 'crisis management and prevention' permits rapid identification of abusive posts and Twitter 'trends', enabling actors to respond quickly to limit the damage. The use of SMS messaging and mobile apps to document GBV, such as USAID's CommCare program and InformaCam, has improved the accuracy of data. In some cases, these mobile tools also assist victims by providing them with referral services via SMS, as is the case with HarassMap. In some countries, Ushahidi has been used to establish an early warning system for election violence. In 2010, in Burundi, a coalition of NGOs – Amatora Mu Mahoro – analysed election violence trends reported via Ushahidi for early warning purposes (Bardall 2010b). Likewise, the Women's Situation Room initiative deployed in several Sub-Saharan African states in recent years (UN Women 2012) has used incident reports collected via SMS, cell phones and other ICTs to provide rapid response for victims (Bardall 2010a). The use of ICTs in the Women's Situation Rooms empowers women to act as stewards of the peace. Through mobilisation, mediation and multi-sector coordination, these programs reinforce this key civic role performed by women in many countries.

A final component of research and documentation is the development of online platforms to store and share knowledge on the issue of ICTs and gender violence. GenderIT.org has been the leader in this area and provides an information resource and knowledge-sharing site for gender and ICT advocates, civil society organisations and policy makers, focused on Africa, the Asia-Pacific, Central Eastern Europe and Latin America. Linked with the Association for Progressive

Communications, under the auspices of the Women's Networking Support Programme, the website promotes issue papers and research exploring the intersection between the internet and VAW, women's rights, sexuality and sexual rights.[18]

ICTs also promote the prevention and mitigation of VAWP in a second area: awareness raising. The connection between election violence and gender-based violence is poorly understood. Training for women candidates and aspirants helps identify the links and empower women to protect and prepare themselves for attacks by using social media tools to respond effectively. Training programmes offered by non-profit organisations like the National Democratic Institute (Borovsky, Cherrabi el Alaoui and Doherty 2010) as well as public resources that orient women aspirants help women use social media to their advantage, fight attacks, establish a credible online image to decrease their vulnerability to attacks and quickly respond to and defend themselves against attack.

ICTs are being used to educate women about other ICT-based risks in ways that can be adapted to responding to VAWP. The International Knowledge Network of Women in Politics (iKNOW Politics) uses a technology-based forum to offer an interactive and multilingual tool that allows members and users to access resources, share expertise and create knowledge through mediated discussion forums and consolidated expert responses to queries. Today, iKNOW Politics offers the most publicly accessible and extensive collection of resources on the issue of VAWP, including country case studies, news, interviews, academic articles and policy papers. Apart from iKNOW Politics, other examples include specific online courses that promote victim safety from ICT-related VAW (see, for example, Safety Net Canada) or offer service providers, anti-violence workers, law enforcement and members of civil and criminal justice systems information about the use of technology to stalk and harass victims (see Western Education Learning Network). Awareness raising targeted at media providers is also necessary, as much of the violence is perpetrated through media channels. Awareness raising and professional standards training for journalists and media professionals are common features of international electoral assistance programs and can provide opportunities to integrate information about VAWP. Finally, ICTs offer creative opportunities to respond to VAWP, for example, through the use of digital storytelling which, by bearing witness to experiences of violence, provides a voice to victims and promotes awareness and action (Roland 2006). Digital storytelling has been used to support victims of GBV and to promote awareness worldwide in the past decade through the work of diverse initiatives[19] as well as through short-form storytelling micro-blogging.[20]

Finally, ICT offers new platforms for advocacy to respond to specific events or in support of individual activists or candidates. ICT advocacy campaigns are especially effective in addressing the cross-cutting issue of VAWP. In particular, NGOs have been shown to effectively use ICTs to further initiatives, raise awareness, forge networks and exchange information on broad issues related to VAW (Hamm 2001). In 2016, the NDI launched the #NotTheCost campaign, mobilising digital advocacy to raise awareness and prevent VAWP. Working in the context of the 16 Days of Activism Against Gender Violence Campaign, the Take Back the Tech campaign seeks to train ICT users in employing activism targeted at violence against women. The campaign draws on an extensive array of ICT and social media tools to empower women at both the personal and broader public levels. These tools include using internet platforms for advocacy; mapping attacks (such as hacking, blocking and deletion) of the websites of women's rights organisations, sexual rights advocates, feminist activists and bloggers; and user-friendly games to promote safety in social networks. The campaign is active in more than 25 countries worldwide. Although the political dimension of VAW is not fully integrated into the campaign, Take Back the Tech is a model for ICT-based advocacy against VAW and therefore offers a platform from which to consider more innovative and impactful future strategies to challenge VAW across a diverse range of contexts. Social media, notably Facebook, was used to mobilize millions of women in an estimated 650+ cities worldwide for the Women's March in January 2017.

Just as they are uniquely able to amplify VAWP, ICTs also empower advocacy around VAWP to help resist it. Speaking together with a united voice empowers victims to affirm themselves, overcome shame and fear and resist the damaging impacts of these acts that seek to silence or coerce their political expression. Awareness and documentary evidence, alongside personal empowerment, are essential to raising the profile of VAWP and reducing such violence by confronting patriarchal norms and pushing cultural barriers that allow it to occur. Nonetheless, notable limits exist to ICTs' ability to respond to VAWP. Monitoring and mapping all forms of VAW are notoriously challenging, and measuring the impact of online attacks requires innovations in research methodology. Awareness raising and empowerment are only possible where ICT penetration and culture are sufficiently deep. ICT-based solutions to resisting VAWP are only complements to legal and institutional measures to prevent and punish VAWP.

Conclusion

ICTs have had a profound impact on the reach and shape of violence against women in elections and politics, creating new threats and obstacles to achieving gender equality in political life. One of the greatest advantages offered by ICT usage to date has been the ability to collect and document incidents of GBEV/GBPV, thereby enabling recognition of the problem and establishing baselines for progress. However, these innovations have a long way to go to catch up to the threats posed by social media–based violence against women in elections. To do so, it is necessary to address the underlying dangers presented by social media – specifically, psychological forms of violence designed to attack women's dignity, morality and self-worth. Both gender and political rights advocates and practitioners seeking to prevent and mitigate this unique form of violence will gain by integrating the best practices from their mutual fields.

Notes

1 This chapter is presented with the generous consent of *Stability: International Journal of Security and Development*, where an earlier version of the text appeared in a research note in 2013 (see Bardall 2013).

2 Reflects single and lower houses. In 1995, women comprised more than 30 per cent of seats in only Sweden, Norway, Finland, Denmark and the Netherlands. As of June 2016, 46 states have met or surpassed 30 per cent women in single or lower-house seats.

3 Also referred to as gender-motivated political violence.

4 Adapted from CEDAW General Recommendation No. 19 on VAW (GR 19).

5 The idea of 'women in politics' is sometimes interpreted broadly to include many categories of women stakeholders, including voters and party/candidate supporters (NDI 2016; ParlAmericas 2014; Tariq and Bardall 2016; UNW 2014), while some authors intentionally use 'in politics' to limit the pool to female candidates and elected or appointed officials (British Group-IPU 2013; Krook 2015; Krook and Restrepo Sanín, 2014). Others describe the objects of VAWIP as including the male and female family members of female candidates/office holders, but exclude female voters, party supporters and so on (IDEA 2008; SAP 2006, 2010).

6 Some view VAWIP exclusively as a form of VAW (Krook 2015; Krook and Restrepo-Sanin 2014, 2016), others view it as a subset of election violence (IDEA 2008) and still others view it as both (NDI 2015; SAP 2006, 2010). Similar concepts face identical debates: for example, ParlAmericas views 'political harassment and violence against women' as a form of VAW, while USAID suggests that 'political violence against women in elections' is a form of election violence.

7 Some authors suggest that motives are exclusively gender based (Krook 2015; Krook and Restrepo-Sanin 2014, 2016 ['with some ambiguity']), while others argue that VAWIP may be driven by both gender-specific and political motives (Hubbard 2015; NDI 2015; Piscopo 2015; SAP 2006, 2010).

8 Empirical research suggests that candidates are one of the least frequent targets – more than 90 per cent of recorded incidents against women in elections were against women in roles other than candidates (Bardall 2011, 2016). The same research indicates that political supporters are the largest category of victims and that female voters face a proportionally much greater risk than do their male counterparts.

9 Men were victims in more than 90 per cent of cases in the Bardall (2011, 2016) studies, and women were involved as perpetrators at the same rate as they were victims. Political violence and GBPV are also targeted at other gender identities.

10 The objective of VAWIP has been variously described as to compel women to 'step down as candidates or resign a particular political office' (Krook and Restrepo-Sanin 2016), to 'prevent and discourage' (IDEA 2008), 'breach, obstruct or curtail' (SAP 2010) and 'prevent or control' (NDI 2015) women's participation. This objective is inconsistent: under some definitions, the goal is to reduce or eliminate women's voices. In other cases, where women's participation can be controlled (by violence, family voting, etc.), increasing women in office can be valuable strategy (Bardall 2014).

11 A debate exists as to whether this form of violence has arisen as a backlash to women's increasing participation (Krook 2015) or is a long-standing issue that has been overlooked to date (Bardall 2016).

12 The tests comprise data from three countries in South-East Asia (Timor-Leste, Nepal and Bangladesh) and two Sub-Saharan African states (Burundi and Guinea) as well as a small dataset from Guyana in South America.

13 See WHO@, a volunteer organisation founded in 1997 to fight online harassment. Statistics are based on incidents reported primarily from the United States in 2012, as well as a limited number of cases in Europe. See www.haltabuse.org/about/about.shtml.

14 Developing-world women are 25 per cent less likely to have access, while women in Sub-Saharan Africa are 45 per cent less likely – see Intel Corporation and Dalberg Global Development Advisors.

15 Author's conversations with women MPs and candidates, including in Tunisia, July 2013; Haiti, April 2013; Nigeria, January 2013; and Egypt, November–January 2011–12.

16 Inversely, sexualised and/or pornographic Photoshopped images of political candidates, short animations distributed online and YouTube promotional videos have also been used to depict the virility of male candidates/officials, such as US presidential

candidate Donald Trump (2016) and Vladimir Putin ('Strip for Putin/Virgins for Putin' 2012) or use sex and violence to encourage electoral participation (see, for example, Danish 'Voteman' 2014).

17 Bolivia's Law against Harassment and Political Violence against Women was passed on 12 May 2012. In 2013, the Mexican senate established a definition of acts that constitute political gender violence.

18 See www.genderit.org.

19 Silence Speaks, the Saartjie Baartman Centre for Women and Children (South Africa), Sonke Gender Justice Network (South Africa).

20 See, for example, #RedMyLips, #YesAllWomen, #OrangeUrHood, and #CarryThatWeight or the #notokay/@KellyOxford Twitter activity in October 2016 around sexual assault testimonies and the US presidential election.

References

Al Jazeera (2013), 'Does Facebook have a "violence against women" problem?', 22 May, available at http://stream.aljazeera.com/story/201305222040–0022771.

Arya, D (2013), 'Why are Indian women being attacked on social media?', *BBC Hindi*, Delhi, 7 May, available at www.bbc.co.uk/news/world-asia-india-22378366.

Association for Progressive Communications [APC] (2013), *The Impact of ICTs on Women's Public and Political Life: Women's Rights Programme*, UN Working Group on Discrimination Against Women in Law and Public Life Meeting.

Baker, L, Campbell, M and Barreto, E (2013), 'Understanding technology-related violence against women: types of violence and women's experiences, learning', *Network Brief* (6), Learning Network, Centre for Research and Education on Violence Against Women and Children, London, Ontario, available at www.learningtoendabuse.ca/learningnetwork/network-areas/technology.

Ballington, J (2016), UN Women Policy Advisor, Political Participation UN Women. 'Re: Definitions – violence against women in elections', Message to G. Bardall, F Tariq, 11 April, Email.

Banks, K (2001), 'Leave the internet alone', *APWIN*, vol. 3, pp. 147–73.

Bardall, G (2010a), *Preventing Election Violence in Burundi*, IFES, Washington, DC, available at www.ifes.org/Content/Publications/Opinions/2010/Apr/Preventing-Election-Violence-in-Burundi.aspx.

Bardall, G (2010b), 'Election violence monitoring and the use of new communications technologies', *Democracy and Society*, vol. 7, no. 2 (Spring 2010), pp. 9–12.

Bardall, G (2011), *Breaking the Mold: Understanding Gender and Election Violence*, IFES, Washington, DC.

Bardall, G (2013), 'Gender specific election violence: the role of information and communication technologies', *Stability: International Journal of Security and Development*, X(X), 1–11, available at www.stabilityjournal.org/article/view/sta.cs/161.

Bardall, G (2014), 'Women's political equality and hybrid regimes', Panel Presentation, *International Political Science Association Annual Conference*, July, Montreal.

Bardall, G (2016), 'Gender and election violence: gender-based distinctions and motivations in political conflict', *9th Annual Women in International Security Canada Workshop*, Mount Saint Vincent University, Halifax, NS.

Blackman-Woods, R (2013), Parliamentary Event organized by the Inter-Parliamentary Union and UN Women on the occasion of the 57th Session of the Commission on the Status of Women on 5 March 2013, UN Headquarters, New York.

Bocij, P (2004), *Cyberstalking: Harassment in the Internet Age and How to Protect Your Family*, Greenwood Publishing Group, Westport, CT.

Borovsky, G, Cherrabi el Alaoui, S and Doherty, M (2010), *Confiance, Compétences, Réseautage: Un Guide Destiné à Toute Jeune Femme Souhaitant Devenir Leader*, National Democratic Institute, National Democratic Institute. Washington, DC.

Breen-Smyth, M (ed.) (2016), *The Ashgate Research Companion to Political Violence*, Routledge, Burlington, VT.

British Group – Inter Parliamentary Union (2013), 'Violence against women in politics', Lecture, Westminster, available at http://unwomenuk.org/violence-against-women-in-politics/

CEDAW General Recommendation No. 19 on VAW (GR 19).

Chaio, R (2011), Harass Map. Interview with Gabrielle Bardall, November 2011, Cairo, Egypt.

Citron, DK (2014), *Hate Crimes in Cyberspace*, Harvard University Press, Cambridge, MA.

Commission on the Status of Women (2013), *Agreed Conclusions on the Elimination and Prevention of All Forms of Violence Against Women and Girls*, available at www.un.org/womenwatch/daw/csw/csw57/CSW57_Agreed_Conclusions_(CSW_report_excerpt).pdf.

Davidson, H (2013), 'Facebook locks out campaigner against images of violence against women', *The Guardian*, 31 May, available at www.theguardian.com/world/2013/may/31/facebook-locks-out-campaigner-women.

Döing, L (2013), 'British banknote prompts hazing on social media', *Deutsch Welle*, 4 August, available at www.dw.de/british-banknote-prompts-hazing-on-social-media/a-16996762.

Ellison, L and Akdeniz, Y (1998), 'Cyber-stalking: the regulation of harassment on the internet', *Criminal Law Review*, December Special Edition: Crime, Criminal Justice and the Internet, pp. 29–48.

Essof, S (2009), *South Africa: Violence Against Women and Information Communication Technologies*, Association for Progressive Communications (APC).

European Consortium for Political Research (ECPR) (2012), *Standing Group on Political Violence*, available at https://ecpr.eu/default.aspx (accessed 10 May 2016).

Falcone, M (2008), 'Palin's e-mail account hacked', *The New York Times*, 17 September, available at http://thecaucus.blogs.nytimes.com/2008/09/17/palins-e-mail-account-hacked/?_r=0 (accessed 16 October 2013).

Hamm, S (2001), 'Information communication technologies and violence against women', *Development*, vol. 44, no. 3, pp. 36–41.

HarassMap (2013), *Women Under Siege*, available at http://harassmap.org/en/what-we-do/around-the-world/ (accessed 4 June 2013).

Hubbard, C (2015), 'Violence and intimidation against women in elections needs to stop: here's how', Democracy Works, *NDI*. Blog, October, available at www.demworks.org/violence-and-intimidation-against-women-elections-needs-stop-here-s-how (accessed December 2015).

Hughes, L (2016), 'Labour MP Jess Phillips receives more than 600 rape threats in one night', *The Telegraph*, 31 May, available at www.telegraph.co.uk/news/2016/05/31/labour-mp-receives-more-than-600-rape-threats-in-one-night/ (accessed 16 June 2016).

iKNOWPolitics (2007), *E-Discussion: Eliminating Violence Against Women in Politics*, 10–14 December, available at http://iknowpolitics.org/sites/default/files/evawip-ediscussion-iknowpolitics2007-summary.pdf.

Intel Corporation and Dalberg Global Development Advisors (2013), *Women and the Web: Bridging the Internet Gap and Creating New Global Opportunities in Low and Middle-Income Countries*, available at www.intel.com/content/www/us/en/technology-in-education/women-in-the- web.html (accessed 9 July 2016).

International IDEA (2008), *Combating Violence Against Women in Politics in South Asia*, available at www.idea.int/asia_pacific/violence_against_women.cfm (accessed 13 July 2012).

InterParliamentary Union (2016), 'Sexism, harassment and violence against women parliamentarians', *Issues Brief*, October, Geneva.

InterParliamentary Union (2016), *Women in Parliaments Database*, available at www.ipu.org/wmn-e/classif.htm (accessed 29 June 2016).

Jamieson, KH and Dunn, J (2008), *The 'B' Word in Traditional News and on the Web*, Neiman Reports.

Kee, Jac sm (2005), 'Cultivating violence through technology? Exploring the connections between Information Communication Technologies (ICT) and Violence Against Women (VAW)', *APC WNSP Issues Paper*.

Krook, ML (2015), 'Empowerment versus backlash: gender quotas and critical mass theory', *Politics, Groups, and Identities*, vol. 3, no. 1, pp. 184–8.

Krook, ML and Restrepo Sanin, J (2014), 'Violence against women in politics: concepts and policy solutions', *American Political Science Association Annual Meeting*, 28–31 August, Washington, DC.

Krook, ML and Restrepo-Sanin, J (2016), 'Gender and political violence in Latin America', *Política y gobierno*, vol. 23, no. 1, pp. 125–157.

Madanda, A, Ngolobe, B and Amuriat, GZ (2009), *Uganda: Violence Against Women and Information and Communication Technologies*, Association for Progressive Communications (APC), Melville, South Africa.

Moolman, J (2013), *The Need to Prioritise Violence Against Women Online*, Report on Women's rights, gender and ICTs, Association for Progressive Communications (APC).

Munyua, A, Mureithi, M and Githaiga, G (2010), 'Women and cybercrime in Kenya: the dark side of ICTS', Working document, 20 July, available at www.genderit.org/resources/women-and-cybercrime-kenya-dark-side-icts-working-document (accessed 16 October 2013).

National Democratic Institute [NDI] (2015), *NDI Launches Global Campaign to Address Violence Against Women in Elections*, May, available at www.ndi.org/violence_against_women_elections (accessed 7 May 2015).

National Democratic Institute [NDI] (2016), *#NotTheCost Campaign*, available at www.ndi.org/not-the-cost (accessed 16 June 2016).

ParlAmericas (2014), 'Political harassment and violence against women', available at www.parlamericas.org/en/group-of-women/political-violence.aspx (accessed 24 March 2015).

Piscopo, J (2015), 'The challenges with legislation as enforcement: rethinking responses to violence against women in politics', Extended version of presentation delivered at the *International Seminar, Violence in Politics against women in Latin America: Diagnostics, Dialogues, and Strategies*, held in Mexico City 11–13 November 2015.

Roland, C (2006), 'Digital stories in the classroom', *School Art*, vol. 105, no. 7, p. 26.

SAP International (2006), *VAWIP Surveillance System: A Guidebook*, SAP International, Kathmandu.

SAP International (2010), *Violence Against Women in Politics: Defining Terminologies and Concepts*, available at www.ndi.org/files/VAWIP_Defining%20TERMINOLOGY%20AND%20CONCEPTS_Final.pdf.

Southworth, C, Finn, Jerry, Dawson, Shawndell, Fraser, Cynthia, Tucker, Sarah (2007), 'Intimate partner violence, technology, and stalking', *Violence Against Women*, vol. 13, no. 8, pp. 842–56.

Take Back the Tech (2016), *16 Days of ICT Action*, available at www.takebackthetech.net/page/about-campaign.

Tariq, F and Bardall, G (2016), *Creating an Enabling Environment for Women's Political Participation: Ending Violence against Women in Political and Electoral Processes*, UN Women, Pakistan.

United Nations [UN] (1995), 'Beijing declaration and platform of action', adopted at the *Fourth World Conference on Women, Strategic Objective J.1.27*, available at www.refworld.org/docid/3dde04324.html (accessed 16 October 2013).

UN General Assembly (2006), *In-Depth Study on All Forms of Violence Against Women: Report of the Secretary-General*, 6 July, A/61/122/Add.1, available at www.refworld.org/docid/484e58702.html (accessed 17 October 2013).

UN Women (2012), *Women's Situation Room Aims to Prevent Conflict and Ensure Peaceful Elections in Sierra Leone*, UNWOMEN, available at www.unwomen.org/en/news/stories/2012/11/women-s-situation-room-aims-to-prevent-conflict-and-ensure-peaceful-elections-in-sierra-leone/#sthash.XnHVBPNP.dpuf.

UN Women (2014), Violence Against Women in Politics: A Study conducted in India, Nepal and Pakistan. *UNWOMEN*, available at www.unwomen.org/en/digital-library/publications/2014/6/violence-against-women-in-politics.

UN Working Group on Discrimination Against Women in Law and Public Life (2013), *Report of the Working Group on the Issue of Discrimination Against Women in Law and in Practice*, available at www.ohchr.org/Documents/HRBodies/HRCouncil/RegularSession/Session23/A.HRC.23.50_EN.pdf.

USAID (2013), *Best Practices in Electoral Security*, USAID, Washington, DC.

Wanyeki, Muthoni, L (2010), 'Lessons from Kenya: women and the post-election violence', *Standpoint, Feminist Africa*, no. 10, pp. 91–97.

Waters, HR, Hyder, AA, Rajkotia, Y, Basu, S and Butchart, A (2005), 'The costs of interpersonal violence: an international review', *Health Policy*, vol. 73, no. 3, pp. 303–15.

Western Education Learning Network Center for Research & Education on Violence Against Women & Children (2016), *Online Training to Promote Safety for ICT-Related VAW*, available at www.vawlearningnetwork.ca/sites/learningtoendabuse.ca.vawlearningnetwork/files/Online_Training_Promote_Safety.pdf.

Conclusion

Marie Segrave and Laura Vitis

The burgeoning analysis and debate that intersect at the nexus of gender–violence–technology, we think, is at an important juncture. There are critical and important voices whose work is dedicated to naming new forms of gendered, sexualised violence (new either because of the use of technology or because they have been enabled via technology) and offering explanations for lines of responsibility (see, for example, Powell and Henry 2017). However, this collection has sought to look beyond this focus whilst contributing to this broader project of researchers and advocates alike, scrambling to find a language to describe, to find evidence of and to create accountability for these unique and novel manifestations of gendered violence. The work of many in this area is important: it is through diverse, collective, robust research that we can bring to bear a clear critique of what is and ought to be happening. This collection affords not only a moment to consider what we know and what we do not know but also an opportunity to consider where we should go. We argue that this collection lays a challenge for further research whilst reckoning with some particularly messy grey areas.

We consider five areas that are clear platforms for future interrogation. Some of this work is already underway and is exciting, while some of it needs to be underway urgently. By way of framing this discussion, we have sought to contribute to – or at least identify – a platform for ongoing feminist criminological interrogation which develops and shapes new ground and understandings of gender, technology and violence by focusing on the key themes of facilitation and resistance.

Incorporating difference: beyond gender

In a collection such as this, the gaps are always noticeable and noted. There is little work here that accounts for the ways in which specific populations (often, minority populations) rely upon and engage with technology in diverse ways and/or the specific impact of gendered violence upon such groups (by way of example, via attention to sexuality, racial or ethnic diversity and ability) or the ways in which this violence and the subsequent responses both produce and reinforce gendered boundaries and expectations. We think, for example, of those who have limited mobility and/or rely on technology as a tool for communication. In such instances, technology offers a way to engage with the world and to be on a more equal footing with others (that is, more independent), and yet, where technology is relied upon and essential to living, it is also necessarily a point of potential and specific vulnerability. There is a need to consider the differential impact, utilisation and awareness of technology. It is equally true that there are many whose lives do not revolve around social media, who are unaware of, choose not to engage with or do not understand how to navigate various technologies. It is not that trolling and sexual harassment online only matter to those who are actively and publicly online, but we must take care to recognise that people's engagement with technology is as multifaceted as technology is itself.

Moreover, the issues raised within this collection also point to the fact that intersections of technology and violence are not simply about what happens within these spaces but that we must acknowledge that these technologically facilitated spaces enable connection between people, both interpersonally and politically. Thus, in examining this space, we are mindful to interrogate how women are excluded from it, either incidentally through conditions of geopolitical location or intentionally through disciplinary tactics. These practices are at once an extension of practices seen in traditional physical spaces where women experience exclusion (in varying degrees), but we also recognise that there are points of difference within the experience of these spaces, their regulation, and the consequences of women's exclusion. As outlined by Bardall, in certain countries, women's use of technology is monitored heavily, while for others, access to technology is limited (or, in the case of China, controlled by the state) – and such practices collectively deny women access to political and civic engagement. Moreover, as this collection testifies, the use of technology to facilitate violence is not just technologically but also culturally located, with social, political and economic impacts. We should take, for example, recent research into the use of text messages, particularly fabricated text messages, as evidence of infidelity, which are then used to justify so-called honour killings (Begikhani et al. 2010). Such research indicates that any future examination of the role of technology and gender should further incorporate women's experiences within the socioeconomic and cultural context that impacts their position in society, in addition to the nature, prevalence and response to violence perpetrated against them. In a similar vein, future research should engage these questions specifically by looking at the intersections of race, gender and sexuality. The examples provided within this collection largely focus on violence against women by men, as does much of the emerging scholarship into technologically facilitated violence. In this regard, future research that explores the nexus of facilitation and resistance should and will also examine how women are or may be perpetrators of technologically facilitated harm, as well as the intersections of gender, violence and technology in the lives of specific groups of people.

Technology and the diffusion of responsibility

Responsibility in this global moment is a concept that seems best described as something that no one – no government, no corporation, no particular individual – wants to have and something that everyone is equally quick to give to others. Many of the contributions in this collection challenge the notion of responsibility – at the individual, state and corporate levels. And there is an important and growing body of work that interrogates individual responsibility, particularly in the broader narratives related to so-called revenge porn, trolling and sexual harassment online, as noted throughout this collection. We applaud the critical interventions this work is progressing.

We also, however, seek to emphasise two related manifestations of responsibility raised in the collection. The first is the repetition across many forms of violence of the potential and/or actual victim-survivor being held responsible for making use of all the intervention and support technologies developed to 'help' or 'protect' – a point highlighted in particular by Maher et al., drawing on Butler, in relation to family violence. So, too, in social media, much of what is documented in this collection involves women pushing back and making decisions about how to respond to and cope with violence online, in the absence of any entity taking clear and proactive responsibility to create safer digital spaces for all.

The second type of responsibilisation is evident in the way in which the feminist project of eliminating violence against women has enabled protectionist measures and gendered tropes to dominate in place of real reform and has thus been subjected to strong critique. We see this when we look to sexual assault rates and the conviction rates in many countries, but we also see this in the embedding of a discourse of women's vulnerability in relation to violence, which can have a range of impacts – as Milivojevic and Segrave point out in relation to how migrant women become potential victims of trafficking. Across both of these examples of responsibilisation, what is most evident is the intrinsic and persistent logic of women *still* being held responsible for violence directed towards them and also that women are most easily identifiable as victims. We also see some evidence of efforts to push back against this from young men and women alike; but actively interrogating the logic of responsibility in the public discourse and corporate and state responses remains critical.

Recognising that the identification of harm and the response to it require interrogation

A central tenet of the work in this collection is that the various forms of harm and violence – harassment, rape culture, political exclusion and trafficking – are as much culturally located and constructed as they are technological. The denigration of women within the public sphere, the curtailing of their political engagement and the reliance on the familiar tropes of women as innocent victims are evidenced in this collection but have been primary concerns for generations. As a result, responses to these forms of violence and harm must be embedded in an understanding of the ways in which these behaviours are gendered as opposed to merely technological and therefore warranting solutions that are not merely technological.

In this vein, this collection also calls into question how resistance is fostered. Each chapter considers the logic of the response to various forms of violence and harm enabled or enacted via technology. Many further interrogate how technology is being harnessed as a response and challenge the presumption that technology is a prime solution for and means of control of social problems. Maher et al.'s chapter, for example, considers and problematises the potential implementation of apps and websites to aid women in violent intimate partner relationships to seek protection and information. In fact, several chapters problematise 'online' or 'technological' protection mechanisms – such as those of Pavan, and Milivojevic and Segrave – as they identify that the architecture of products and policies aimed at protection can in fact create or enable spaces for harassment and exploitation to proliferate. As such, this illustrates the necessity of continuing to reflect upon the potential implications of the use of technology to expose, prevent and manage technologically facilitated violence rather than relying on the assumption that technology is inherently best positioned to provide solutions.

Research with a broader view of harm

The issues brought to the fore within this collection extend how we understand violence and the new harms engendered via technology. The range of harms considered within this collection is broad. Thus, a central question remains: how do we articulate such harms? The chapters in this collection take different approaches to this question – where Milivojevic and Segrave challenge the articulation of e-trafficking, others such as Jane, and Maher, McCulloch and Fitzgibbon, explore how harassment is embedded within emerging and novel technological spaces that are socially mediated. The latter chapters consider what it means to be mob attacked, to be surveilled via GPS or to constantly receive misogynistic comments on social media questions – issues that are highly specific to an emerging online discourse. The chapters address questions with which not everyone is engaged, specifically those responsible for creating law, making policy and implementing protection efforts. Moreover, ambiguities abound in this domain: what constitutes normative and harmful behaviours online is not definitive, and this is most evident in recent attempts to criminalise online harassment.

The recent legislative reform to criminalise the use of information communication technologies to harass in the United Kingdom (UK) and Canada illustrates such ambiguities. In these two jurisdictions alone, there are indications that the law is being readily applied to instances of online harassment. For example, in 2014, 1,209 people were convicted under section 127(1a) of the UK Communications Act 2003, which criminalises sending a grossly offensive, obscene, indecent or menacing message via a public electronic communication network, compared to only 143 convictions in 2004 (Ministry of Justice 2015). Such numbers may be perceived initially to be indicative of a proactive application of these laws for the benefit of swift policing of online spaces. However, a recent case in Canada has highlighted the complexities involved in online harassment satisfying the threshold of criminal harassment. In a landmark Canadian 'Twitter harassment' case (*R v Elliott* [2016] ONCJ 35), Canadian man George Elliott was charged with harassment for repeatedly tweeting at (via hashtags) two Canadian women after they had blocked him and asked him to stop. The complexities of criminalising harassment on new platforms like Twitter was brought to the fore when Justice Knazan dismissed these charges on the basis that, while the behaviour constituted harassment, it did not cause the complainants to fear for their safety. In Justice Knazan's sentencing decision, he stated that the decision 'to subscribe to Twitter and keep your account open is a decision that 'waive[s] your right to privacy in your tweets', such that Elliot's refusal to cease tweeting at the complainants did not constitute harassment (*R v Elliott* [2016] ONCJ 35, at 54).

This case reflects what has been observed throughout this collection: first, that there is a broad range of harms women are facing within Web 2.0; second, that there have yet to be normative standards articulated with regard to online behaviour and, finally, that the pace of new technologies, the rapidity of their uptake and their pervasiveness in everyday life have given rise to urgent and complex questions about how to theorise techno-social harms and articulate why they impact women in particular. As Pavan notes in her chapter, there are still those who interpret such behaviour as merely 'idiots speaking online'. Yet as Jane argues, online sexual violence 'inflicts many layers of suffering, in that women often suffer not only the initial assault, but victim-blaming and sex-related shaming in the aftermath'.

As researchers, we are faced with the task of continuing to explore and interrogate how and to what extent these practices engender harms but not necessarily with the sole aim of facilitating the criminalisation of behaviour. For scholars working within interdisciplinary fields, online spaces give rise to longstanding questions about how to best articulate gendered harms in ways that are nuanced and theoretically informed. But if legal and regulatory frameworks aimed at tackling technologically facilitated violence are to be improved and developed, research that maps and develops nuanced understandings of harm is essential. We are reminded at this point that much of the research about formal criminal justice responses to technologically facilitated harassment indicates that the key challenges faced by victims are that police do not understand the severity of the incidents involving technology, the technologies through which abuse is fashioned and/or the range of applicable legislation (Citron 2014). As a consequence, their advice to victims may be to simply turn off their computers or switch off social media, by relying upon the long-departed distinction between offline and online spaces. This issue was brought into sharp relief in the recent landmark Australian online harassment case, in which a Sydney man was convicted of using a carriage service to menace, harass or offend after sending 55 sexually explicit and aggressive comments to a woman on Facebook (Hunt 2016). As detailed in the media, the woman who made the complaint reported that her first interactions with police revealed a technological divide: 'The cop I spoke to didn't even have Facebook – explaining to her the post, the reposting, the screenshotting, the comments, was harder than it needed to be' (Lyons et al. 2016, para 57). What this indicates is that a wide range of robust scholarship that examines the implications of technologically facilitated harms is necessary to bridge gaps of understanding between key stakeholders.

The importance of technology in the gender–violence–technology nexus

In some ways, the issues and case studies raised across these chapters collectively point to technology as an add-on, an extenuation of violence that already exists, or an additional tool to react to or prevent gendered violence. Potentially, the focus on technology is a distraction from the deeply embedded historical reality of the relationship between gender and violence in all its forms. To some extent this is true: promoting real change whereby men and women do not accept any form of violence against women and gender equality is realised in politics, labour, the economy and society – these are the real changes that are required and the goals to work towards. However, technology is more than just an add-on. As the chapters in this collection bring to the fore, we must not only consider how technology can be used for good and for bad but also cast a critical gaze over some of the assertions about the role of technology in producing or sustaining some forms of violence. We must also question the growing focus on cybersecurity as *the* technological dilemma of our time – detailed in this collection are social, more complex forms of violence that are equally urgent.

The development of technology and the capacity of violence to be enacted through it are not new phenomena. The impact of firearm technology on the enacting of violence globally and the requisite response to control firearms to reduce this violence is just one example. However, we cannot view technology simply as a 'weapon' nor as merely a threat to be controlled. What is evident across this collection is that technology is an intrinsic part of our lives – it allows politicians to campaign, people to communicate, victim-survivors to seek help anonymously, migrants to access routes and advice on how to travel undocumented from one country to another, couples to be sexually intimate and free speech. However, it also enables these things to be disrupted and fouled: it can enable harm.

A way forward

The examination of what constitutes gendered and sexual violence is increasingly being debated, explored and contested within online spaces. Importantly, discussions about 'what should be done' and what is 'being done' about gender, violence and technology are now taking place within diverse networked spaces, in some cases imbued with a feminist language and grammar. There is a necessity to continue to map and interrogate such movements and to expand the criminological gaze to consider these as legitimate cultural moments in which the discourse around issues like sexual violence is taking place in networked publics, as well as to consider what knowledge, activism, insights and contestations are being generated within these spheres.

What we believe is that the important, growing body of research underway on technologically facilitated violence will continue to unravel what is occurring and what we can do and that there is a parallel body of work that carefully considers the assumptions underlying the broader discourse about the assertion of technology as either a problem or a solution. We offer this collection to the growing body of work in this area, certain that this is an important time for critical feminist scholarship and that the work of many in this area will collectively create change. We remain indebted to the feminist criminologists, activists and advocates who came before us – naming gender and naming violence and refusing to accept simplistic solutions and rhetoric. We need to recognise that this work is occurring in a new world order in which nation-states wield power alongside global corporations and the future of the United Nations seems somewhat weakened by growing isolationist political leadership. We are emboldened by the work within this collection and the work of others broadly in this area. From the moment of inception to the arrival of this collection, much has changed in the world – technologically and politically. This pace will not slow, and our challenge is to be mindful, alert and open.

References

Begikhani, N, Gill, A, Hague, G and Ibraheem, K (2010), *Final Report Honour-Based Violence (HBV) and Honour-Based Killings in Iraqi Kurdistan and in the Kurdish Diaspora in the UK*.

Citron, D (2014), *Hate Crimes in Cyberspace*, Harvard University Press, Cambridge, MA.

Hunt, E (2016), 'Judge compares offensive Facebook posts to football in sentencing Sydney man, *The Guardian*, 28 July, available at www.theguardian.com/australia-news/2016/jul/29/court-sentences-sydney-man-for-sexually-explicit-facebook-comments (accessed 29 July 2016).

Lyons, K, Phillips, T, Walker, S, Henley, J, Parrell, P and Carpenter, M (2016), 'Online abuse: how different countries deal with it', *The Guardian*, 12 April, available at www.theguardian.com/technology/2016/apr/12/online-abuse-how-harrassment-revenge-pornography-different-countries-deal-with-it (accessed 20 April 2016).

Ministry of Justice (2015), *Outcomes by offence tables, Criminal Justice System Statistics Quarterly*, December 2014, available at www.gov.uk/government/statistics/criminal-justice-system-statistics-quarterly- december-2014 (accessed 1 December 2014).

Powell, A and Henry, N (2017), *Sex, Violence and Justice in a Digital Era*, Palgrave Macmillan, Basingstoke.

Legislation

UK Communications Act 2003.

Case law

R v Elliott, 2016 ONCJ 35 (CanLII), available at http://canlii.ca/t/gn1hq (accessed 23 January 2016).

Index